Performance Engineering

Giuseppe Serazzi

Performance Engineering

Learning Through Applications Using JMT

 Springer

Giuseppe Serazzi
Milano, Pavia, Italy

ISBN 978-3-031-36762-5   ISBN 978-3-031-36763-2 (eBook)
https://doi.org/10.1007/978-3-031-36763-2

This Springer imprint is published by the registered company Springer Nature Switzerland AG
The registered company address is: Gewerbestrasse 11, 6330 Cham, Switzerland

In memory of
Larry Dowdy, Martin Reiser, Paul Schweitzer, Kennet Sevcik
before the night falls

Preface

This open-access book aims to improve users' skills needed to implement models for performance evaluation of digital infrastructures. Models are widely used in any branch of engineering. Unfortunately, their use for performance evaluation of computing infrastructures is pretty much limited to modeling specialists and not to their end-users, who have complete knowledge of the analyzed phenomena. Among the reasons for this limitation, there is the intrinsic complexity of the modeling process, which cannot be fully learned with the academic approach alone, and the frequent use of unnecessary mathematical details, which typically create a fog shield that hides the key features of the models. Furthermore, it is known that to increase the ability to build reliable models, it is necessary to accumulate experiences that can only be learned through trial-and-error work by solving problems of different difficulties.

Based on these considerations, we tried to keep this book as simple as possible by adopting the following guidelines. On one side, we present a collection of modeling studies of increasing complexity, describing the assumptions made and motivating the decisions taken. Readers are introduced to the modeling process gradually, learning the basic concepts step-by-step as they go through the case studies. On the other side, we try to avoid superfluous exposure to mathematical concepts. For interested readers, we reported in *Appendix* some basic notions that may be useful to know.

Among the various techniques used in performance evaluation modeling, we will use *Queueing Networks*, possibly integrated with *Petri Nets* when the characteristics of the models require it. This type of model provides a good balance between the accuracy of results, complexity, and parameterization effort, for a large variety of problems. Analytical, simulation, and asymptotic techniques are applied to solve the models.

The book is structured in six Chapters and an Appendix. Chapter 1 is focused on the description of the model building process. The input parameters, the output metrics, and the operational laws are illustrated. The most important steps to building models to be solved with simulation and analytical techniques are reviewed. In the following Chapters, fifteen case studies of increasing complexity covering different aspects of performance evaluation are described. We believe that readers could benefit from analyzing these models by focusing on the abstraction process applied to their design. In Chaps. 2 and 3, models of systems with homogeneous and heterogeneous workloads are presented, respectively. The problem of bottleneck identification and performance optimization is addressed for both types of workloads. Chapter 4 is devoted to the analysis of the impact of variability of the traffic of requests and service demands on throughput and response time. Chapter 5 focuses on parallel computing and describes the influence of different synchronization policies on performance. Chapter 6 presents four case studies derived from real-life scenarios: a surveillance system, an architecture that autoscales for load fluctuations, a web app workflow simulation, and a crowd computing platform. The autoscaler model consists of Queueing Networks and Petri Nets integrated, i.e., it is a *multi-formalism* model.

The `Java Modelling Tools` (`JMT`), a *open source* suite of six tools for performance engineering and capacity planning using Queueing Networks and Petri Nets, were applied to build and solve the models. Details on `JMT`, which can be downloaded from `http://jmt.sourceforge.net`, can be found in [8]. `JMT` is a project coordinated and co-developed by Politecnico di Milano (G. Serazzi) and Imperial College London (G. Casale).

This book is intended as a text for courses in performance evaluation and modeling for graduate and senior-level computer science students. Researchers and practitioners whose work is related to performance evaluation of computer infrastructures will find it useful as a reference text. It can be used also as a supporting text for courses in disciplines outside of computer science that require the use of modeling to evaluate the performance of their applications.

We hope you will find the *learning through applications* approach followed in this book useful for your work, and apologize in advance for the mistakes you will find. The author cannot be considered responsible for errors that you may introduce in your work due to the content of this book.

Milano, Pavia, Italy   Giuseppe Serazzi
May 2023

Acknowledgments

The author wishes to thank Giuliano Casale for his important effort in supporting the JMT project. Thanks to his fundamental coordination and co-development work, the JMT suite has achieved the technical level it is recognized for. *Thanks for your scientific support!* Special thanks also go to Marco Gribaudo, for the enlightening and endless discussions that have increased my knowledge on performance modeling. *Thanks for your patience!* The author is grateful to Andrea Sianesi and Eugenio Gatti, respectively, President and General Director of the Fondazione Politecnico di Milano, for supporting the publication of this book as Open Access. *Thanks for your enthusiasm in accepting my request!*

Giuseppe Serazzi

Contents

Chapter 1
The Process of Modeling

1.1 Model Implementation

Typically building models is a fairly easy task, but making them *accurate representations* of the phenomena to be reproduced is a *completely different* matter. The potential sources of error are so many! Among them, the most frequent regard the wrong interconnections of the components, the inaccurate values assigned to input parameters and the incorrect use of the techniques and tools adopted to implement and solve the models.

The construction of a model requires several iterations at different levels of granularity. At the highest level the user need to iterate many times between *three* different operational environments: the *Real world*, the *Abstract space*, and the *Modeling domain*. The process of modeling is outlined in Fig. 1.1. In this book we will focus primarily on the *Modeling Domain* phase.

A modeling study typically begins in the *real-world environment* with the observation and measurement of the phenomenon that must be reproduced. The interactions among the various components, or resources, must be assessed and their influence on the behavior of the phenomenon must be investigated. The variables that describe the quantities that we measure or we want to estimate will be referred to as *performance metrics* or *indexes*. Usually, some *key* components are very *critical* for the success of the study for the strong influence they exert on the metrics of interest, while others have a negligible impact on them. Since the effect on the results produced by the latter are minimal or zero, they can be ignored without affecting the validity of the model and thus greatly reducing its complexity.

Among the key components of the model it is very important to identify the most requested one, i.e., the *bottleneck*, as it determines the performance of the overall system. The saturation condition of the bottleneck depends on the characteristics of both the service requests and the component itself. Note that due to the typical fluctuations of a workload, both in the intensity and in the amount of work required, the bottleneck can *migrate* among different resources generating abrup changes in performance. Ignoring this condition may invalidate the entire modeling study. Bottleneck

© The Author(s) 2024
G. Serazzi, *Performance Engineering*,
https://doi.org/10.1007/978-3-031-36763-2_1

Fig. 1.1 Operational environments involved in the modeling process

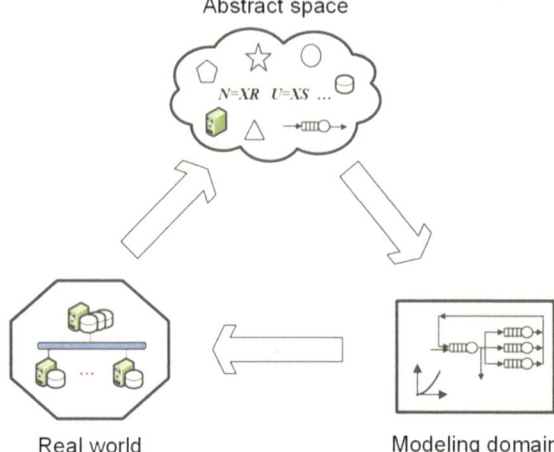

migration is very common when the workload consists of different applications with highly variable service demands.

The technical features of the components, the assumptions introduced, the workload considered, and the statistical properties of the variables used must be accurately described. Depending on the type of study, the data required for these operations can be derived directly from measurements or must be generated by analytic functions.

The knowledge gained from the analysis of the real-world phenomenon will be merged in the *Abstract space* (see Fig. 1.1) with the technical background of the modeler. In this phase, decisions on the modeling technique to be adopted, on the plausibility of the assumptions introduced, on which components should be considered, on the metrics to be used, and many others must be taken. The actions that must be done in the abstract space rely upon the experience and creativity of a person rather than her/his technical background. These skills can neither be learned with an academic approach nor passively absorbed; instead them build up with experience gained through daily trial-and-error work.

The work in the abstract space end with the implementation of a first version of the model. We are now entering in the *modeling domain*. The process of implementing a model is intrinsically *iterative* and an *incremental approach* is usually adopted, and strongly recommended.

The components represented in the model are progressively increased starting from a small initial set that must include the bottleneck and other key components that have a great impact on the results. At each iteration the complexity of the model is increased until the level of detail of the metrics obtained matches that of the objectives of the study.

Once a model with its workload are completely defined and parameterized, the *validation phase* starts to assess its *accuracy*. Several methods of analysis are used to evaluate different properties of the model. The techniques adopted are function of the type of study to be conducted.

A model should be *calibrated*,its *sensitivity* and *robustness* must be assessed, its performance must be forecast with a projection technique, and the domain of validity of its results must be evaluated. Typically, a validation technique requires several *iterations* at low level of granularity. When the system to be analyzed is operating, the performance metrics obtained as model *outputs* must be compared with those *measured* from the real system while it processes the production workload. In this case, the *calibration* of the model consists in the tuning of its input parameters so that the differences between the two sets of performance metrics (the measured outputs compared to those of the model) are negligible or at least tolerable. In case of unacceptable differences, the input parameters must be recomputed and the assumptions introduced (including the layout of the model) should be revised.

Great attention must also be paid to the selection of the measurement interval (the *observation interval*) in which data on the behavior of the system components and the workload executed are detected. The data must be collected when the workload processed is *representative* of the load that is typically executed by the system. In some cases, the measurement period may consist of several *disjoint* intervals. The main steps of the modeling design process and the operations required by the *incremental approach* are shown in Fig. 1.2.

1.2 Inputs and Outputs of Models

Any computing system can be viewed as a set of *resources* (hardware and software) that execute the processing *requests* submitted by users. Therefore, the input parameters of a model can be divided into two groups regarding the *load* and the *resources*, respectively. Depending on the system being modeled, in the following *processing requests* will hereinafter be interchangeably called *jobs*, *applications*, *customers*, *requests* or *users*, and the *resources* will be referred to as *stations*, *elements*, *components*, or *service centers*.

The arriving requests are collectively called *workload*, while *workload characterization* refers to their quantitative description [11]. When the individual workload components have similar characteristics, they are grouped together and their statistical parameters, such as mean, standard deviation, and distribution, are used as inputs to models. In this case, the workload is referred to as *homogeneous* or *single class* and the models are called *single class models*. The components of a workload that consists of various types of applications typically have significantly different service demands. In this case, several groups of components with similar characteristics must be identified and the workload is referred to as *heterogeneous* or *multiple class* (*multiclass*). Each class will be described with its statistical characteristics.

In this Section we focus on the inputs and outputs of *single class* models. With multiple class workloads the notations become more complex (an index for the classes must be added to the identifiers of the stations), but the meaning of the parameters remain the same. Models with single class workloads will be described in Chap. 2 while models with multiclass workloads will be analyzed in Chap. 3.

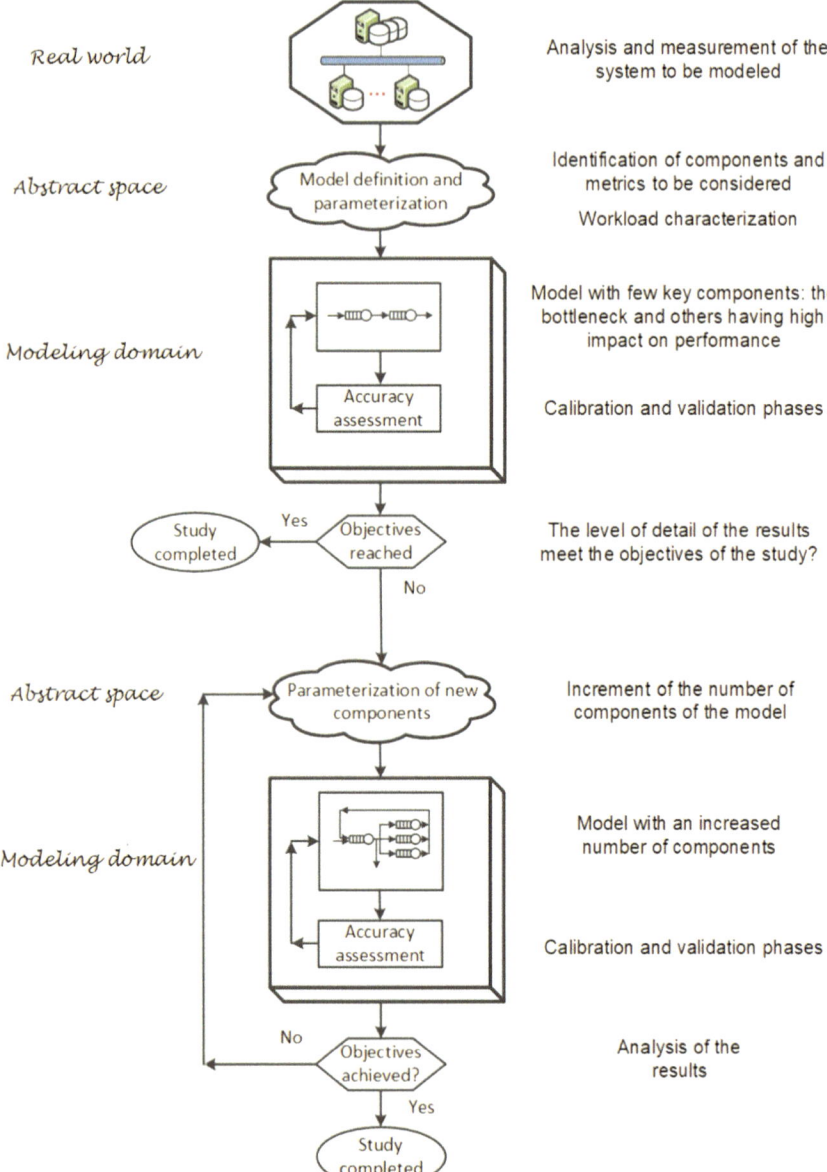

Fig. 1.2   Incremental approach to model building

Table 1.1 Some basic input parameters and output measures of *single class* queueing network models with queue and delay stations

	Component-level (index r)	System-level (index 0)
Input parameters	*type* of component (queue, delay, ...)	model *type*: open, closed
	V_r number of visits per each job execution	K total number of components
	S_r service requirement per visit	λ_0 jobs arrival rate (open models)
	D_r service demand of a job	N_0 number of jobs in syst. (closed models)
	Z_r think time (for delay component)	Reference station
Output measures	X_r throughput of the component	
	N_r number of requests in the component	X_0 system throughput (closed models)
	U_r utilization	N_0 number of jobs in syst. (open models)
	Q_r queue time (for queue component) R_r response time of a request Rd_r residence time of a job	R_0 system response time

Since performance models can be implemented at different levels of detail, the metrics described in Table 1.1 are divided in two groups whether they refer to the component-level or to the system-level. These metrics are the basic ones used in single class models consisting of queue and delay stations. In a queue station requests compete for the server and wait in queue to receive service, then leave the station when finished. In a delay station there is no server competition as it is assumed that there is always a free server for all incoming requests.

In *single class* models, the class index can be avoided. When a metric refer to a single component of the model, the subscript identify the specific element considered. The subscript 0 (zero) is used to identify metrics which refer to the system as a whole. For clarity, whenever possible, we will refer to the service requests arriving at each station in a component-level model as simply *requests*. In system-level models, the computational requests submitted by the users will be identified as *jobs* or *customers* interchangeably.

A description of some basic *input parameters* and *output results* for models that use queue and delay stations follows. The algebraic relationships among some of them, derived in [10, 16, 26], are also reported. Metrics for other types of stations, e.g., fork/join, finite capacity regions, Petri Nets place/transitions, will be described in the case studies where they are used.

Throughout this book we have tried to keep the description as general as possible, but since we have used the JMT Java Modelling Tools to solve the models, it has sometimes been necessary to refer to terms specific to the individual tools used, i.e., JSIMg (*the Simulator*), JMVA (*the Analytical solver*), and JABA (*the Asymptotic Bound Analyzer*).

The open source JMT suite can be downloaded from http://jmt.sourceforge.net.

Let us remark that a large part of the queueing networks solved with analytical techniques are of *separable* type. This subset of general queueing networks can be solved analytically with very efficient algorithms. Clearly, the property of being *separable* introduces some restrictions to the system characteristics that can be modeled. Some of them concern the *concurrent* use of resources, the *constraints* on the number of requests, the *adaptivity* of the routing, the *priority* scheduling algorithms, the *blocking policies*, the *creation and deletion* of jobs, the *dynamic change* of Service times (see Sect.2.3 and, e.g., [4, 25]).

Often these limitations have a minimum impact on the behavior of the system modeled and can be bypassed easily changing the assumptions adopted. In some cases, system characteristics that cannot be modeled directly with separable networks may have a negligible influence on performance. In other cases, the global model can be splitted in several sub-models, some separable and other not, that may be solved with different techniques. The results obtained from the solution of the individual sub-models can then be combined with various techniques in order to obtain the solution of the original global model.

In any case, it is important to note that these limitations *only affect models that are solved analytically*, while those solved with *simulation are not (or minimally) affected*.

Input Parameters

Open/Closed (types of workloads, types of models)
Workloads, like models, may be of two types: *open* or *closed*. When the workload is open, the number of customers in the model is fluctuating and can grow indefinitely as a station becomes saturated, while with a closed workload this number is kept constant. An example of an open workload is the flow of requests arriving from a Internet connection, which is usually quite bursty. The models that execute open workloads are referred to as *open models* while those executing closed workloads are called *closed models* (see Fig. 1.3).

An example of a closed workload is a computing infrastructure that can only be accessed by employees of a company. The number of customers is fixed and limited to the company employees. When the maximum number of customers that can be in execution simultaneously is reached, a new customer can enter the system only when a customer completes its execution. In simulation models with open workloads the customers arriving to the system are generated by a Source station, and at the end of the execution are routed to a Sink station (in JSIMg also Fork, Class Switch, and Transition stations may generate customers). Models with both types of open and closed workloads running concurrently are also possible, and are referred to as *mixed* models.

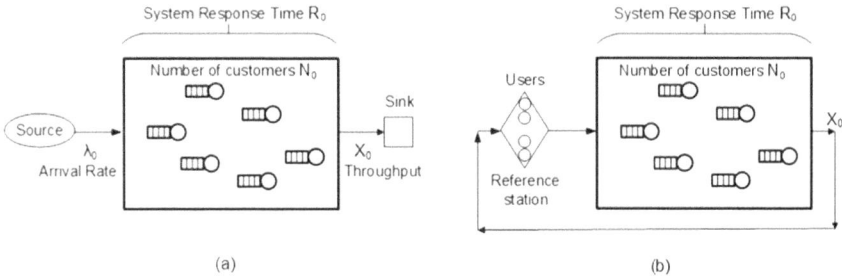

Fig. 1.3 Examples of the two basic types of models: *open* (**a**) and *closed* (**b**)

λ_0**–Interarrival times** (workload intensity in *open* models)
Describes the characteristics of the incoming flow of requests arriving at the system. In analytical models (JMVA), the exponential distribution of Interarrival times is typically assumed as default, and in this case only the *arrival rate* λ_0 is required. In simulation models (with JSIMg), different types of distributions (e.g., burst, hyper-exponential, Erlang, Pareto, etc.) may be selected and some related statistical indexes should be defined. The number of these parameters varies as a function of the distribution.

N_0**–Number of customers** in the model (workload intensity in *closed* models)
This parameter in closed models refers to the mean number N_0 of customers in the model. A job arrives at the system, circulate among the service stations (the resources of the system), and then departs at the completion of the execution. In closed models it is immediately replaced by a new job with the same characteristics, keeping N_0 constant.

Type of component: Queue, Delay, Source, Sink, Class Switch, Fork/ Join, Semaphore, Place/Transition, Finite Cap. Region, Router
In a model, the components representing system resources can be of different types. The most used components in analytically solved models are typically of two types: queue and delay. In queue components, requests arrive, compete for the server, wait in the queue when it is busy, execute when it becomes idle, then exit. A delay component simply introduces a delay in the flow of requests, but no queue is created. In this case, an arriving request will always find an idle server since they are assumed to be infinite. Many more types of components are used in simulation models depending of the tool considered and the complexity of the system to be modeled. For example, in JSIMg Fork and Join stations are used to simulate parallelism, FCR Finite Capacity Regions to control access to model regions, Semaphore to block selectively the requests, Place and Transition to simulate Petri Nets.

V_r–Number of visits per job (to each component)
During its execution, a job visit the components (CPU, disks, storage, ...) several
times before leaving the system. V_r is the mean number of visits, also referred to as
visit count, that a job makes to station r during its complete execution.

S_r–Service time per visit (for *queue* components)
The mean time required to component r to execute *one* service request, corresponding to *one visit*, is referred to as `Service time`. This value does not include the
time waiting in queue when the server is busy. The mean value of S_r and/or other statistical indexes (distribution, variance, coefficient of variation, etc.) must be provided
according to the technique adopted to solve the model (analytical or simulation). See
also comments made above for `Interarrival times`.

D_r–Service demand per job (for *each* component)–$D_r = V_r\ S_r\ D_r = B_r\ /\ C$.
The global amount of service time required by a job to component r for a complete
execution is referred to as `Service demand` D_r. Its value is given by the product
of the `Service time` S_r required by a visit to the component r by the `Number
of visits` V_r that the job makes to it. The D_r are important because it can be
shown that the solution of most queueing networks does not depend on the single values of V_r and S_r but *only* on their product, i.e., *only* the service demand matters. This
property, exhibited by *separable* queueing networks, is important as its application
reduce the number of input parameters and the complexity of the models. Furthermore, the D_r can be measured more easily than V_r and S_r since, very often, their
values are stored directly in the system log files. Another possibility to obtain the
values of the D_r is to divide the total busy time B_r of component r by the number of
jobs C completed in the observation interval. The high level of aggregation adopted
in the models that use the D_r makes it impossible to compute the `Throughput` and
`Response time` at the single component level, while the performance indexes at
the system level are obtained correctly. A more detailed description of `Service
demands` and *separable* networks can be found in the Case Study Sect. 2.3.

Z–Think time per visit (for *delay* components)
In a delay component, the requests never wait in queue since a server is always
available for their execution which therefore always takes on average Z time units
(this time is typically referred to as `Think time`). Z is the correspondent of service
time S of queue components and can be considered as the mean delay introduced by
a delay component to the flow of requests that goes through it. A *delay* component
is often used to represent the *users* in closed models. For this reason it is commonly
selected *by default* as `Reference station` of a model.

Reference station for each workload class (at system level)

The station used to compute the performance indexes at the system level (throughput, response time, global utilization of resources, etc.) is referred to as `Reference station` (RS). When a job flows through an RS, in most cases it is implicitly assumed that its execution has completed and therefore it is leaving the system. In this case, a job visits the RS only once during its life and for this reason the delay station that in most models simulates the users is often selected as RS. However, any of the components of the model may be selected as RS. Clearly, all the performance indexes are affected by this choice. To compute their values, the visits to each component of the model must be scaled with respect to those made to RS. In *open models* JSIMg assumes by default a `Source` station as RS. When a job completes its execution, its performance indexes are computed considering the time interval elapsed between its generation from the `Source` and its exit from the `Sink`. In *closed models* any station can be selected as RS. In this case its performance indexes are calculated considering the time elapsed between its generation from RS and enters the model and the time it exits the model and reaches the RS (see Fig. 1.3).

Output Measures

X_0**–System Throughput** (*at system* level)–$X_0 = N_0/(R_0 + Z) closed model$
$\lambda_0 = N_0/R_0$ *open model*   *Little Law*

This metric represents the *rate* at which the *jobs* complete their executions and leave the system. To compute this metric it is fundamental to know which is the *Reference station* of the model since only the jobs that visit it have completed their executions. Without loss of generality, the visit to the external part of the model is often assumed to be one. X_0 may be computed applying *Little law* to the system as a whole (see Fig. 1.3).

N_0**–Number of jobs** (at *system* level)–$N_0 = X_0(R_0 + Z)$ *closed model*
$N_0 = \lambda_0 R_0$ *open model*

In open models the number of jobs in the system is an output metric since it depends on the arrival rate and on the contention of the components. When the utilization of a component is close to one it saturates and the value of N_0 grows to infinity.

R_0**–System Response Time** (at system level)–$R_0 = (N_0/X_0) - Z$ *closed model*
$R_0 = N_0/\lambda_0$ *open model*

The amount of time required by a complete execution of a job is referred to as `System Response Time`. In closed models, it can be seen as the time interval between two consecutive visits to the `Reference station` by the same job (the first corresponds to the instant of time in which the job is generated and the second to the instant of its completion). In open models, R_0 corresponds to the time interval between the generation of a job by the `Source` station and the moment in which it completes its execution and reaches the `Sink` station. Since the models are in *equilibrium* and each job entering the system visits the `Sink` usually once, the `Source` can be considered as `Reference station`.

X_r–Throughput of component r–$X_r = X_0 V_r$ *Forced Flow law*
Number of requests processed in a time unit by the component r. Note that the unit of measure is *requests* per unit of time and not *jobs* per unit of time (used by System Throughput). When the Service demands D_r instead of the Visits V_r and Service times S_r are used as input parameters, the X_r cannot be obtained from the model. In this case, the throughput of each component is equal to X_0. The relationship between X_0 and X_r are obtained from the *Forced Flow law*.

N_r–Number of requests in component r
In a queue component, this metric refers to *all* the requests in the station, whether waiting in queue or in execution. In a delay component all the requests in the station are in execution, thus N_r corresponds to the mean number of requests in service.

U_r–Utilization of component r–$U_r = X_r S_r = X_0 V_r S_r = X_0 D_r$ *Utilization law*
Fraction of time the server of a *queue* component r is busy (in a station with one server). In a *delay* component this value corresponds to the mean number of requests in service.

Q_r–Queue time of component r (per request)–$Q_r = R_r - S_r$
Mean time spent in queue waiting for the server in a *queue* component.

R_r–Response time of component r (per request)–$R_r = N_r / X_r$
Mean time required to execute the processing request of *one* visit to component r. Its value includes all time spent in the component during a visit, whether waiting in queue or being served. The unit of measure is the time per request. If the number of servers of the *queue* component is one, then it will be: $R_r = Q_r + S_r$.

Rd_r–Residence time of component r (per job)–$Rd_r = V_r R_r$
The *total time* spent by a job at component r during its *complete execution* (including both the time spent in queue and the time being served) is referred to as Residence time Rd_r. While the Response time R_r is *local* to a component (i.e., it may be computed considering *only* the Response times of *one visit* to the resource), to compute the Residence time Rd_r of a resource it is necessary to know the Number of visits V_r that a job makes to the resource *during its complete execution* (see Appendix A.1). The unit of measure is *job*. The System Response Time is the sum of the Residence times of all the components of the model.

1.3   Parameterization of Simulation Models

The sequence of operations required to implement a model is clearly influenced by the techniques and tools used. In this section we restrict our attention to the *simulation* which, compared to other techniques, allows for maximum generality in terms of system architectures that can be modeled and adoptable assumptions.

In the following description of the steps required to implement a simulation model (see Fig. 1.4) we tried to be as general as possible. However, since the models were solved with the JSIMg simulator, its characteristics clearly influenced the sequence of operations performed. The figures mentioned in the flowchart of Fig. 1.4 show some user-interface windows for setting the parameters required by the steps represented.

Implementing a model begins with describing its components and their interconnections. Depending on the types of user interface available, a graphical (see, for example, the model of Fig. 1.5 created with JSIMg), or other type (e.g., wizard) of description can be done.

The parameters for the *workload characterization* are: *type* of customers classes (open or closed), arrival rate and distribution of interarrival times (in some cases also other statistical parameters are required) for *open* classes, number of customers for *closed* classes, and Reference station. In Fig. 1.6 the following parameters for Class1 are set: open class, arrival rate $\lambda = 1$ req/s, exponential distribution of interarrival times, and the station Source1 as Reference station. To select different arrival rates or distributions, simply click on the Edit button and the list of available distributions will be shown. With multiclass workloads the parameters of each class must be provided.

The next step is setting the *station parameters* for all workload classes. In JSIMg, the parameters for a Queue station are organized into *three sections*: Queue, Service, and Routing. In the Queue Section, the Capacity size (max number of customers allowed in the station, in queue and in service), the type of scheduling algorithm, the queue policy, and the Drop Rule (in stations with *limited* capacity) are set. In the Service Section, the Number of Servers of the station, the type of service Strategy whether *load independent* or *load dependent*, and the statistical parameters of the Service Times Distribution are set. In the Routing Section the routing strategies of jobs on the interconnections automatically detected among the stations may be described. For example, in Fig. 1.7, for station Queue1 the Probability has been set as Routing Strategy, and the customers in output are sent to Queue2 with probability 0.3 and to Queue3 with probability 0.7.

The next step concerns the selection of the metrics (*performance indexes*) that must be computed with the model. Usually, for each metric, several statistical variables must be set. In JSIMg (see Fig. 1.8) the following parameters are required: the *class* of *customers* and the *station* (or the *entire system*) considered, the *confidence level* (see Appendix A.2 and, e.g., [36, 37]), the *maximum relative error*, and the decision whether to generate the file with *all the collected values* of the metric analyzed or not. In Fig. 1.8 five indexes concerning Class1 customers are selected: two aggregated at *system level* (System Response Time and System Number of Customers), and three at the Queue1 station level: (Response time, Number of customers, and Utilization). For the Response time, the generation of the CSV file with all the values of the samples analyzed is required (i.e., the Stat.Res checkbox is *flagged*). The 99% confidence level (default value) is required for all the indexes and the max relative error tolerated is

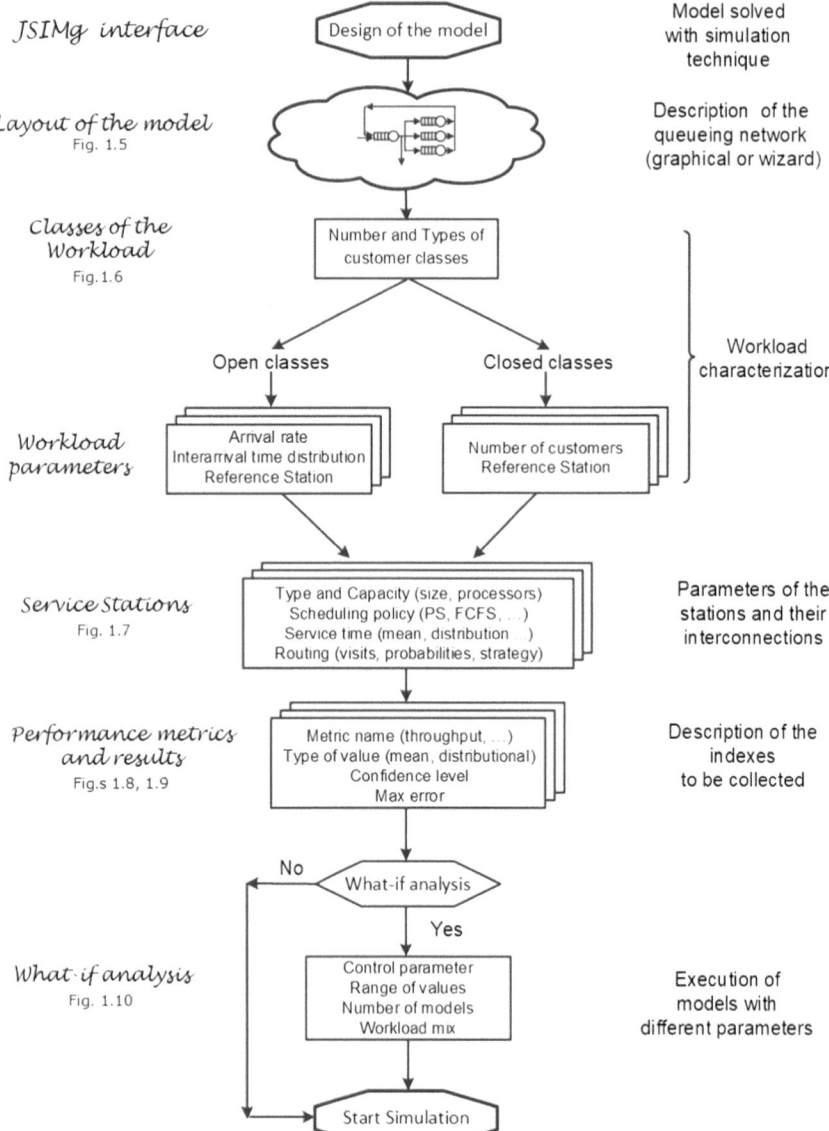

Fig. 1.4 Main steps to implement a simulation model with JSIMg. Figure numbers in the flowchart refer to examples of the corresponding screenshots

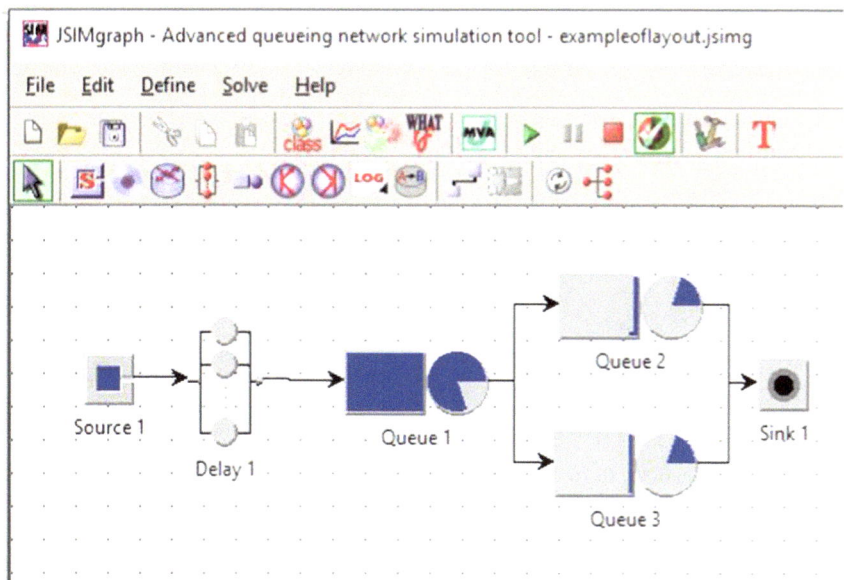

Fig. 1.5   Example of the graphical representation of a model using JSIMg

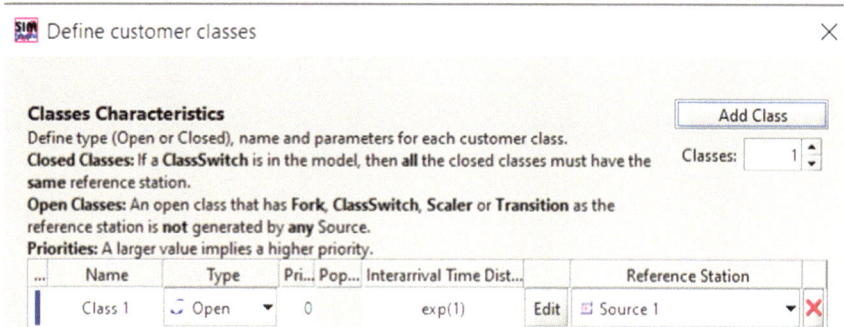

Fig. 1.6   Definition of the parameters for the open class `Class1`

0.03. The simulator *no longer collects data* of an index when the required accuracy is achieved.

When the simulation starts, for each selected index a graph like the one of Fig. 1.9 is plotted. As the simulation progress, the behavior of the confidence intervals and of the mean value of each index are shown together with the number of samples analyzed. According to the request (see the `Stat.Res.` checkbox flagged in Fig. 1.8), the CSV file was generated with *all* the values of `Response times` and the statistical indexes were computed. The CSV file will contain, among the other variables, the values of the *percentiles* (see the example of Figs. 2.10, 2.11).

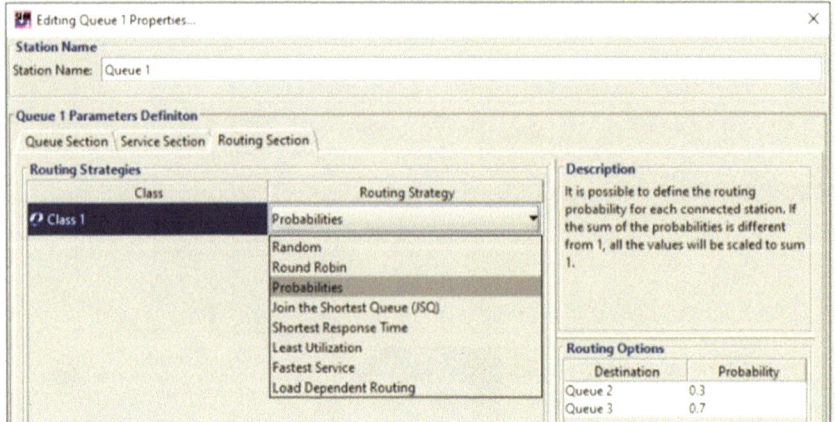

Fig. 1.7 Parameters of the `Routing Section` of `Queue1` station

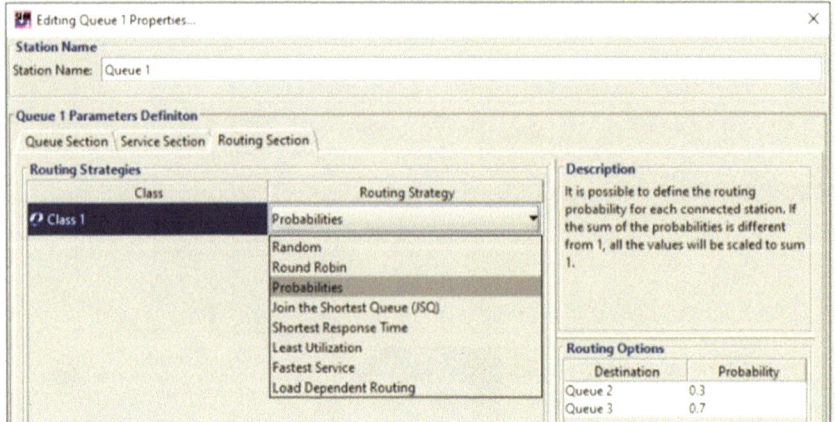

Fig. 1.8 Performance indexes to be collected, their precision and statistical requirements

Most of the performance studies require the evaluation of the impact on system performance of one or more parameters. To meet this objective it is necessary to execute a sequence of models increasing (or decreasing) at each step the value of a parameter, e.g., `Arrival rate` or `Number of customers`, referred to as `Control parameter`. To make this process efficient, many simulators show a feature called `What-if`. For example, Fig. 1.10 shows the `Response times` obtained from the execution of 10 models with `Class1` customer `Arrival rate` increasing from 0.2 to 1.2 job/s. Mean values and confidence intervals are also reported.

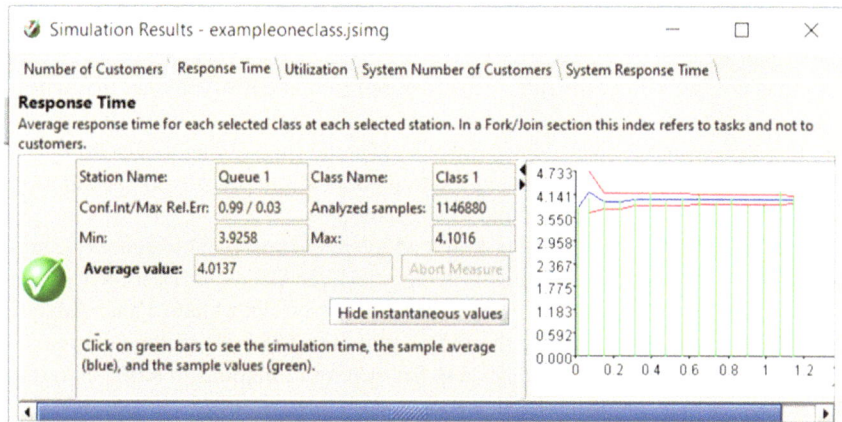

Fig. 1.9 `Response times` of `Queue1` station: mean value and confidence intervals computed on different samples collected during the simulation progress

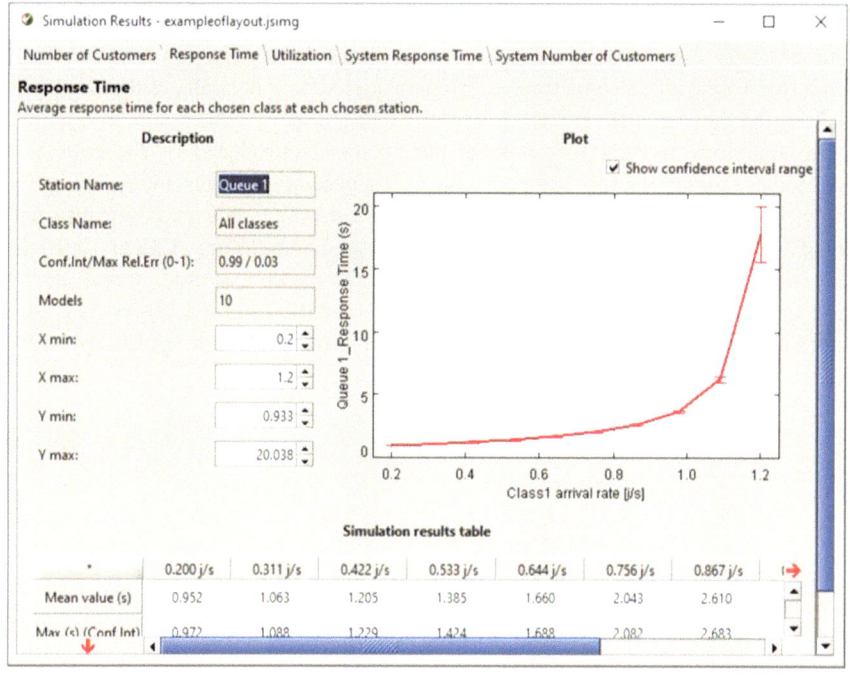

Fig. 1.10 `Response times` of `Queue1` station as a function of the `Arrival rates`

1.4 Parameterization of Analytical Models

In this section we will outline the steps required to implement a model that will be solved with a analytical technique. As can be seen from Fig. 1.11, these steps are not very different from the ones already described for the simulation models in Fig. 1.4. Clearly, the analytical technique and the tool adopted introduce some peculiarities on the operations that can't be found in simulation.

As a function of the solution algorithm adopted, the analytical techniques can be subdivided in exact, approximate, and asymptotic. Each technique has its own parameters. In the following, we will refer to models solved with Mean Value Analysis (MVA) technique [25, 31] using the JMVA tool. The MVA algorithm compute the exact values of performance indexes, but has several limitations in terms of system characteristics that can be modeled.

In the screenshot of Fig. 1.12 the MVA has been selected as solution algorithm for the closed class Class1 of 10 customers. The tabs are reported with the sequence that must be followed for the parameter settings: Classes, Stations, Service times, Visits, Reference station, What-if. In Fig. 1.13 the mean values of the Service times of the three stations CPU, Storage1, and Storage2 are set. As requested by the MVA algorithm, the values of these parameters are considered exponentially distributed. The visits to the three stations are $V_{CPU} = 10000$, $V_{Storage1} = 5499$, $V_{Storage2} = 4500$.

In simulation models, to minimize the overhead introduced by the collection of the data, users should select only those performance indexes interested in the study. In analytical models, however, a consistent set of indexes is always computed as their derivation is very fast (see, e.g., Fig. 1.14). In Figs. 1.15 and 1.16 the Utilizations and the Residence times of the three stations are plotted for the Number of customers ranging from 10 to 100 (90 models were executed with a What-if analysis). The values of the performance indexes are also provided in tabular form.

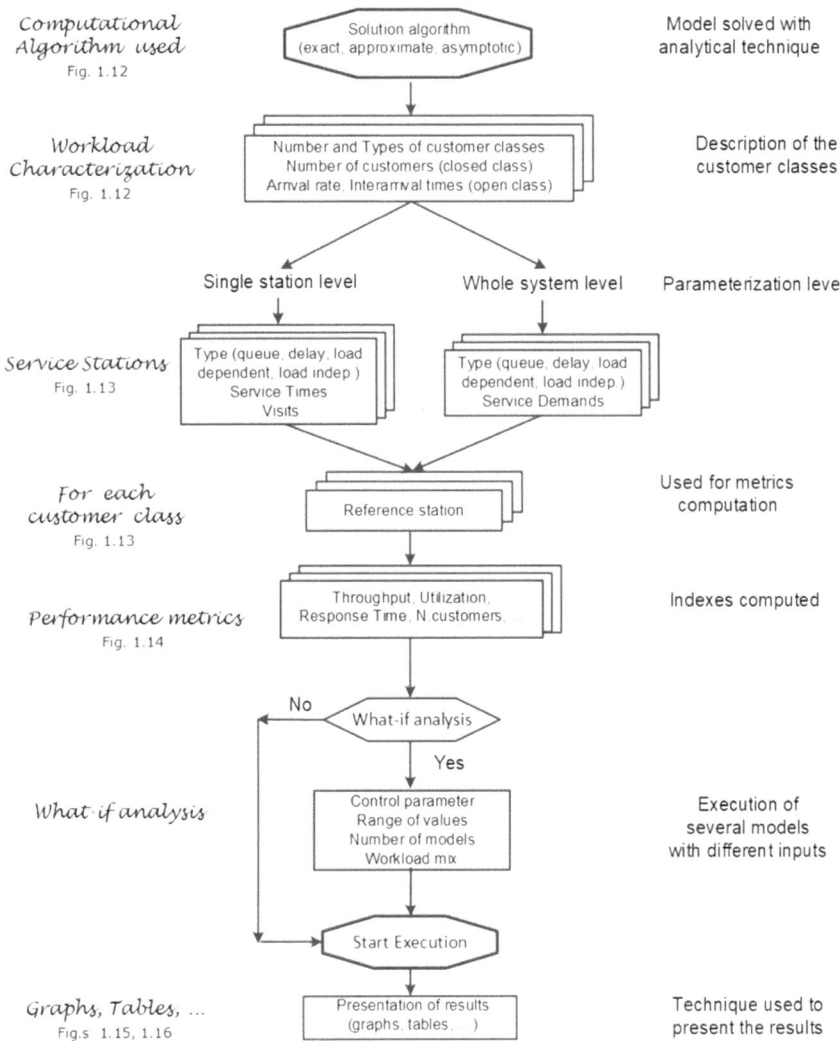

Fig. 1.11 Main steps for the implementation of a JMVA model solved with analytical technique. Figure numbers in the flowchart refer to examples of the corresponding screenshots

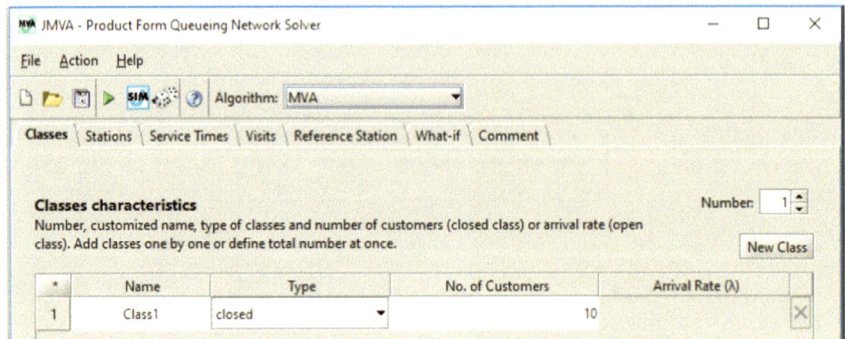

Fig. 1.12 Selection of the MVA solution algorithm and settings of one closed class `Class1` with `10 customers`

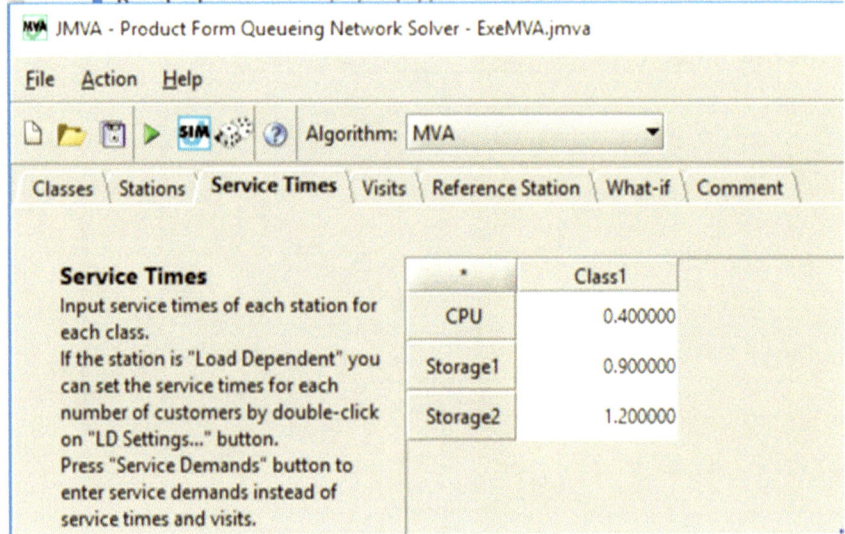

Fig. 1.13 Settings of the `Service times` of the three stations

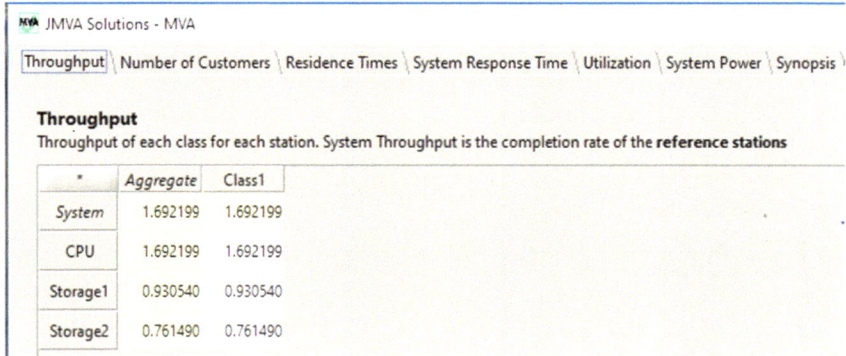

Fig. 1.14 `Throughput` of the three stations. All the performance indexes are computed

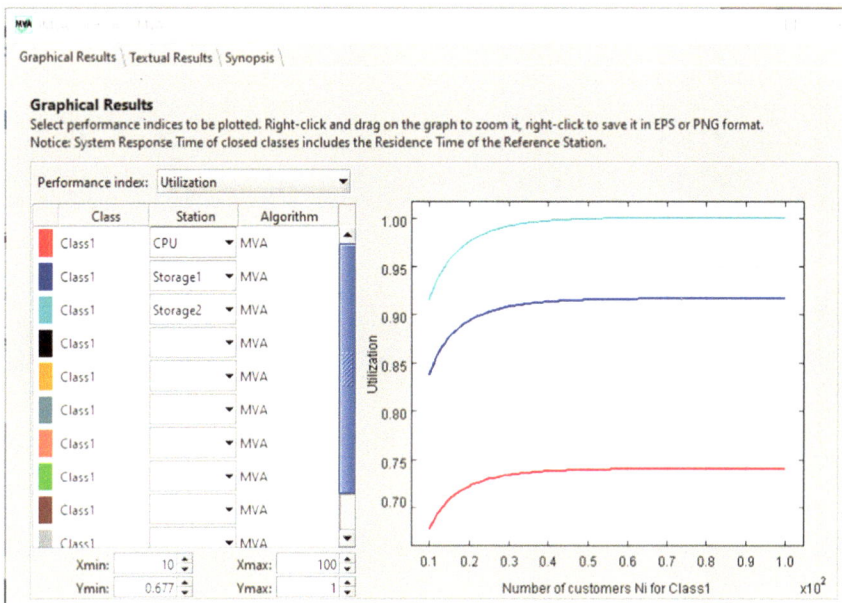

Fig. 1.15 `Utilizations` of the three stations as a function of the `Number of customers`

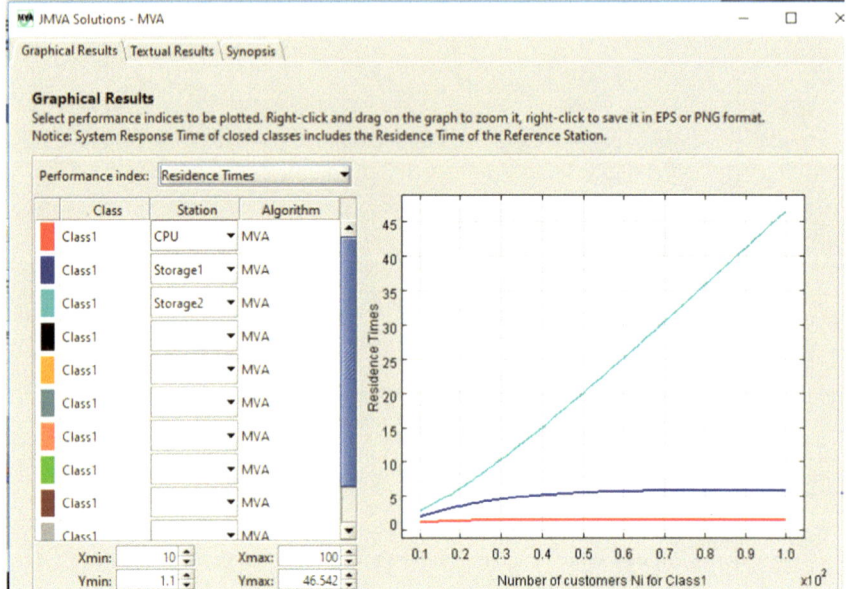

Fig. 1.16 Residence times of the three station as a function of the Number of customers

Chapter 2
Systems with Homogeneous Workloads

2.1 A Web Server with External Workload

tags: open, single class, Source/Queue/Sink, JMVA.

The models for analyzing the performance of a system can be developed at different levels of detail and with a single element that can represent the system as a whole or just one of its components. In spite of their high level of aggregation, models in which the resources of a system are *collectively* represented with a *single component* (i.e., the system is modeled as a black box) yields in many cases interesting results. These models can also provide useful insights for the evaluation of more complex scenarios.

To solve the model presented in this section we use the analytical tool JMVA that applies the classical Queuing Networks equations.

2.1.1 Problem Description

A capacity planning study is required to model a web server utilized for the distribution of technical documentation concerning the products of a company and accessible by a high number of users through Internet. Requests arrive at the server from the network, compete for the resources, and once executed leave the system, see Fig. 2.1a. These models are usually referred to as *open models*. The workload consists of a *single request class*. The requests have similar service demands, are independent each other and arrive to the server with exponentially distributed interarrival times. We consider a simple high-level aggregated model, i.e., a single queue station, representing the web server accessed by a request only once before leaving the station, see Fig. 2.1b. This single-station model may seem inadequate to describe a web server that has at least two resources, a CPU and a storage, that are visited many times by the requests during their execution.

© The Author(s) 2024
G. Serazzi, *Performance Engineering*,
https://doi.org/10.1007/978-3-031-36763-2_2

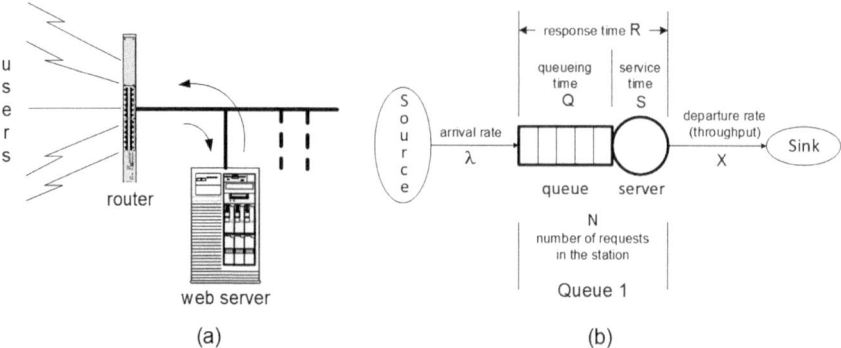

Fig. 2.1 Web server model at high level of abstraction

However, there are several problems in which single-station models yield interesting results. Among the motivations that make them useful in many situations are:

- it is common the case in which only one resource of a system is the dominant concern for the performance objectives, while the remainder components have a negligible impact on them. Modeling this critical resource clearly provides useful information about overall system performance.
- usually, one resource is much more utilized than the others (i.e., it is the *bottleneck*) and is largely responsible for the overall system performance. Models with only this resource can provide accurate predictions of the overall system performance.
- a technique to implement large models is to partition them in smaller submodels and to study them in isolation. The solutions of the submodels are then combined in order to analyze their impact on the behavior of the global model. The station used to represent collectively the stations of a single submodel is called FES, *Flow Equivalent Server*. The objective of a FES station is to introduce in the flow of requests the same delay as the submodel it represents (see, e.g., Chap. 8 of [25]).

The assumption of *single class workload* is important in many situations for the accuracy of the models. When the workload components have significant differences in resource requirements, i.e., when there are *multiple class* requests, the bottleneck may migrate among resources as a function of the fluctuations of the *mix of requests* in execution (see, e.g., [2, 3]). The effect of this migration may be dramatic for the accuracy of the results. With single class workloads the bottleneck does not switch among resources provided that all of them are *load independent*, i.e., their service time is not a function of the number of requests that are in the resource (waiting in queue and in service).

Concerning the single visit hypothesis, this should not be a concern. Depending on the abstraction level of the model, it may not be necessary to explicitly describe the load of each component r of a system at the lower levels of detail using the `Service times` S_r and the `Visits` V_r but it is sufficient to consider the *global* `Service` *demand D_r* placed on each of them during a complete execution (i.e., $D_r = V_r S_r$). To reduce the number of parameters and the effort required by their measurement, we will parameterize most of the models with the `Service demands` D_r.

2.1.2 Model Implementation

We consider a simple *open model* at high level of abstraction, i.e., system-level, (see Fig. 2.1b) consisting of a single *queue* station `Queue 1`, representing the web server that is accessed by the requests generated by the `Source`. Once executed, the requests leave the *queue* station for the `Sink`. In these *open models* the number of requests in execution is *not controlled* by the system itself but depends on the characteristics of the traffic generated by the `Source` (rate and fluctuations of arrivals, service requirements). Depending on these parameters, a system can be flooded with requests whose number can suddenly grow to very high values.

Requests are assumed to be independent of each other and arrive at the server at random times. This is equivalent to saying that `Interarrival times` are *exponentially* distributed. All the requests are considered statistically equal, i.e., are indistinguishable each other, and leave the server at random times. The randomness of the departure times has as the consequence that the `Service times` S, i.e., the time requirement per visit, are *exponentially* distributed.

All requests arriving at the station can be accepted for execution, i.e., there is *enough space* to store them all that can grow indefinitely. This type of station is usually referred to as M/M/1 station (see, e.g., [36, 37]). The arrival rate is $\lambda = 0.2$ req/s, thus the average `Interarrival time` is $1/\lambda$. The average time required by a complete execution of a request is 1 s. This time usually is referred to as `Service demand` D of a request, but since the number of visits to the server in the aggregated model is one, its value coincides with the `Service time` S. Thus, for simplicity, we will use the notation S instead of D in this example. The requests are served according to their order of arrival, i.e., with a FCFS scheduling.

With the hypotheses considered, this model can be solved analytically with the classical *Queueing Networks* equations implemented in the JMVA.

2.1.3 Results

In what follows we will describe the operations required to achieve some of the **objectives** (referred to as *Obj.1–Obj.4*) of the capacity planning study.

Obj.1: implement a model of the server and compute the performance indexes with the parameters above described
In Fig. 2.2 the input parameters for the model solved with QN (Queueing Networks) equations are shown. Some of the performance indexes computed by the model are

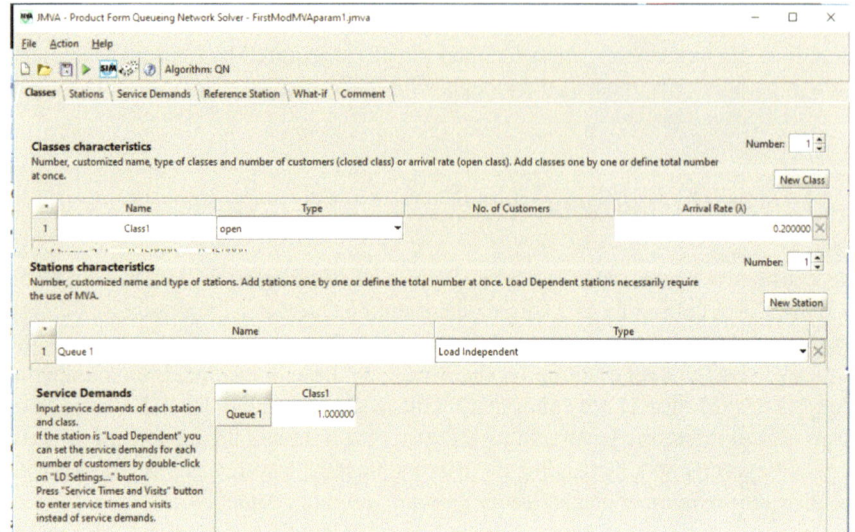

Fig. 2.2 Input parameters of the JMVA for the open model of Fig. 2.1b

shown in Fig. 2.3. The mean number of requests N in the server is 0.25 *req* and the mean Response time is 1.25 s. To check the correctness of the results we computed the values of the same indexes with the exact equations of Queueing Networks:

$$N = \frac{U}{1-U} = \frac{\lambda S}{1-\lambda S} = 0.25 \, req. \qquad R = \frac{S}{1-U} = \frac{S}{1-\lambda S} = 1.25 \text{ s} \quad (2.1)$$

Obj.2: compute the behavior of the performance indexes when the workload increases to $\lambda = 0.9$ req/s.

A What-if analysis is required with Arrival rate as Control Parameter ranging from 0.2 to 0.9 req/s. In Fig. 2.4 the parameterization of the What-if (100 models are requested) and the behavior of two performance indexes, i.e., the Throughput X and the Response time R, are shown. Since in the model there is only the Queue1 station, its Throughput and Response time coincide with the ones of the System. The linear behavior of the Throughput X is correct since we increase linearly the Arrival rate λ from 0.2 to 0.9 req/s and the models are in equilibrium, i.e., it is $\lambda = X$. The *maximum* Throughput of the server corresponds to the *saturation load* $\lambda^{sat} = 1/S = 1$ req/s.

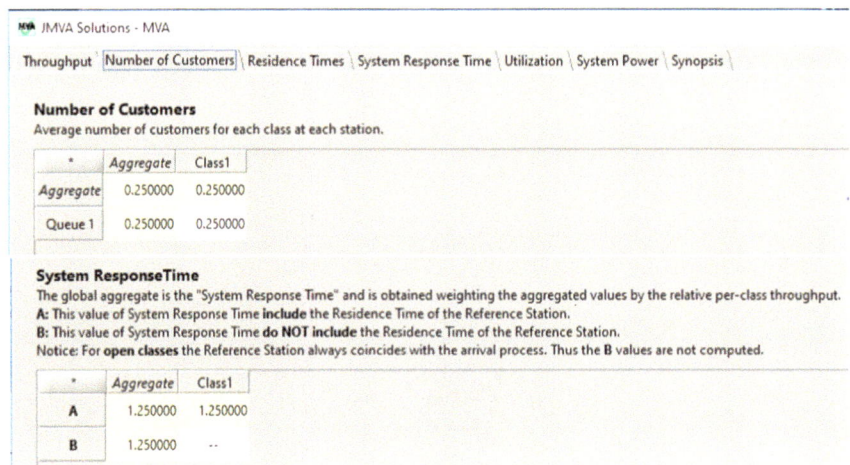

Fig. 2.3 Some of the performance indexes computed by the JMVA

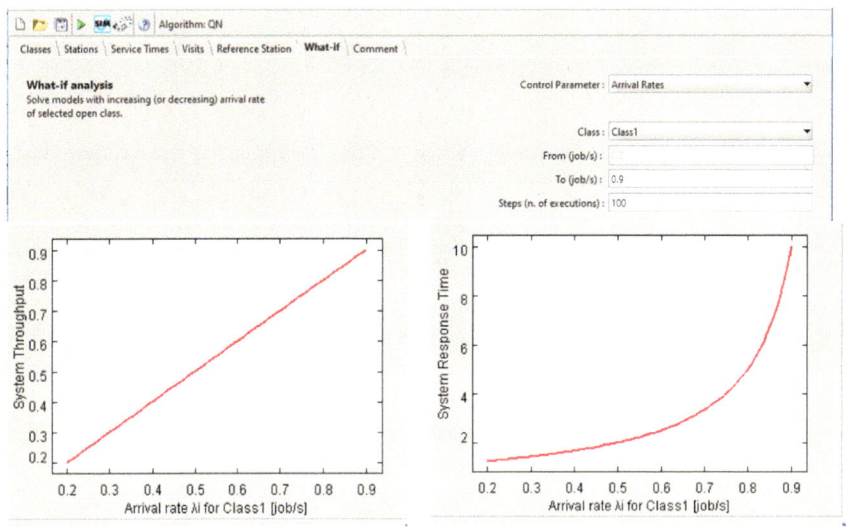

Fig. 2.4 `System Throughput` and `System Response Time` of 100 models with `Arrival rates` λ ranging from 0.2 and 0.9 req/s obtained with a `What-if` analysis

Fig. 2.5 `What-if` analysis: tabular results of the `System Response Times` corresponding to the `Arrival rates` $\lambda = 0.793$ and 0.801 req/s

Obj.3: according to the marketing department forecast, a maximum `Response time R = 5` s can be tolerated. Compute the maximum increase of the workload that satisfy this constraint (% with respect to the original $\lambda = 0.2$ req/s).

We can use the results provided in *tabular* format from the `What-if` analysis made in the previous step. Figure 2.5 shows the `Response times` of models 85 and 86 that are just above and just below the value $R = 5$ s (4.85 and 5.02, respectively). The `arrival rates` used in the two models are 0.793 and 0.801 respectively. So we are sure that $\lambda = 0.793$ req/s satisfy the constraint. Just as a simple check, we may use Eq. 2.1 to derive the value of λ corresponding to R = 5 s and S = 1 s. We obtain $\lambda = 0.8$ req/s. Thus, the increment of the workload *tolerated* is 300%.

Obj.4: a new set of complex technical manuals are expected in the near future whose `Service demand` is assumed to be S = 2 s. What will be the `System Response Time` R with an expected arrival rate double the actual one (i.e., $\lambda = 0.4$ req/s)?

New values for the input parameters S = 2 s and $\lambda = 0.4$ req/s must be set. The corresponding value of `System Response Time` is R = 10 s.

2.2 A Computing Infrastructure with a Closed Workload

tags: closed, single class, Delay/Queue, JSIMg.

In this section we describe a model of a computing infrastructure with a *closed* workload (see Sect. 1.2) solved with the simulation technique. The main characteristic of this type of workload is that the number of customers in execution is constant. A new customer enter the system when a customer complete its execution.

On the basis of the assumptions made, this model could also be solved analytically with JMVA. However, we have used the simulation technique to provide a first simple example of implementing a model with a simulator. Furthermore, it should be noted that simulation is by far the most popular modeling technique used in performance engineering. Indeed, *simulators are very powerful tools* and the set of models they can implement is practically unlimited given the great generality offered in terms of characteristic of the systems and type of assumptions that can be represented.

2.2.1 Problem Description

A computing infrastructure, located in a large data center, is used to execute applications that are very critical to the company's business. This infrastructure adopts very high security techniques to control accesses that are reserved *only* to a *limited* number of authorized employees. It mainly consists of three servers: a Web Server (WS) and two servers (AS_1 and AS_2) dedicated to the Application and Storage functions, see Fig. 2.6a.

Due to the apps executed, the resource requirements of the user requests are similar, i.e., the workload is *single-class*. The Service times of the three servers have different mean values, and are assumed exponentially distributed. The probabilities (i.e., the *routing probabilities*) that the requests in output from the Web Server are routed to servers AS_1 and AS_2 are known. In some problems, instead of the routing probabilities, the *visits* that a request perform to each resource during its execution

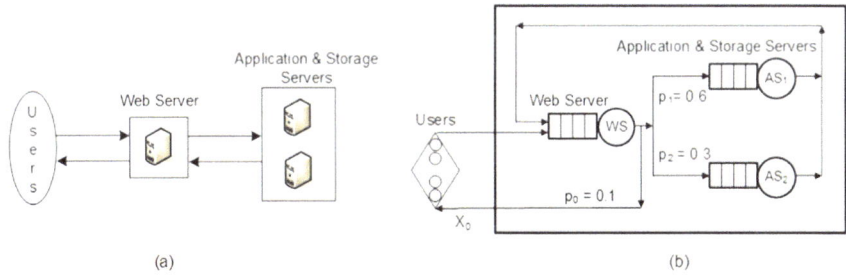

Fig. 2.6 The computing infrastructure considered (**a**) and the corresponding queueing network (**b**)

are known. These two sets of values are related each other and to derive one set from the other it is required to know the *topology* of the network. In Appendix A.1 it is described how to obtain the relationships between the *routing probabilities* and the *visits* for the topology considered in Fig. 2.6b. Assuming that when a request leaves the model it has been completely executed, i.e., that it is $V_0 = 1$, we have:

$$V_{WS} = \frac{1}{p_0} = 10 \qquad V_{AS_1} = \frac{p_1}{p_0} = 6 \qquad V_{AS_2} = \frac{p_2}{p_0} = 3 \qquad (2.2)$$

Models can be parameterized with one set of values or the other. JSIMg accept *both* types of parameters. The scheduling algorithm adopted by the resources is FCFS.

2.2.2 Model Implementation

Since the *number of users* (i.e., the employees authorized to access the computing infrastructure) is *constant*, we implement a *closed model* with four stations: one *delay* and three *queue*, see Fig. 2.6b. Each user submit one request. The probabilities p_i's that after a visit to the Web server WS a request is routed to App&Storage servers AS_i are known. The index 0 is used to represent the world outside the system, and the metrics with index 0 are at *system-level*. Therefore, X_0 and R_0 represent the Throughput and the Response time of the *global system*, and p_0 is the probability that a request leaves the system as it has completed its execution. We assume that a request is routed to this path only *once* in his lifetime, so the number of visits V_0 that it performs outside the system is one. According to the layout of the model it is $\sum_{i=0}^{2} p_i = 1$.

The workload is generated by a station external to the system representing the Users, that we consider as Reference station. This station is used to compute the System Response Time R_0 and the System Throughput X_0. R_0 is defined as the period of time between the instant in which a request enters the model (leaving the Reference station) and the one in which it leaves the model (entering the Reference station). X_0 is the rate of completed requests that leave the model and enter the Reference station. Others performance indexes are also influenced by the selection of the station that will be considered as reference (see Appendix A.1). The mean Service time for each Visit to servers WS, AS1 and AS2 are: $S_{WS} = 0.005$ s, $S_{AS1} = 0.020$ s, and $S_{AS2} = 0.025$ s, respectively. The *think time* of the delay station Users is $Z = 1$ s. All the values are exponentially distributed. The JSIMg model of Fig. 2.7 was solved with simulation. The *routing probabilities* of the requests leaving the Web Server are: $p_0 = 0.1$, $p_1 = 0.6$, and $p_2 = 0.3$, see Fig. 2.8.

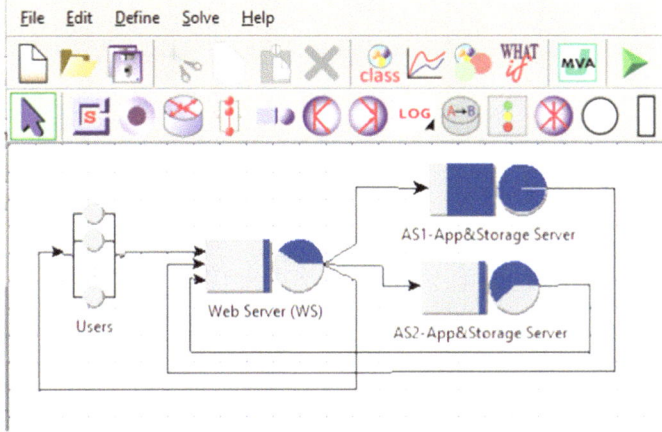

Fig. 2.7 The JSIMg model of the computing infrastrucure of Fig. 2.6b

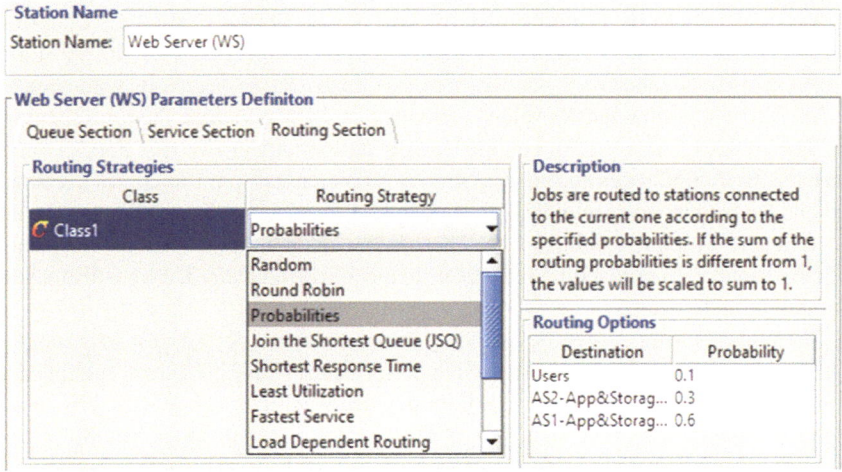

Fig. 2.8 Settings of the Routing Probabilities of the Web Server WS

2.2.3 Results

Several objectives of the capacity planning study were set. In what follows we will describe the results of some of them referred to as *Obj.1–Obj.4.*

Obj.1: Implement the model of the computing infrastructure with the parameters assigned. Investigate the behavior of System Throughput X_0 and System Response Time R_0 for the Number of Customers N_0 ranging from 1 to 20. Which will be the 90th percentile of R_0 with $N_0 = 20$?

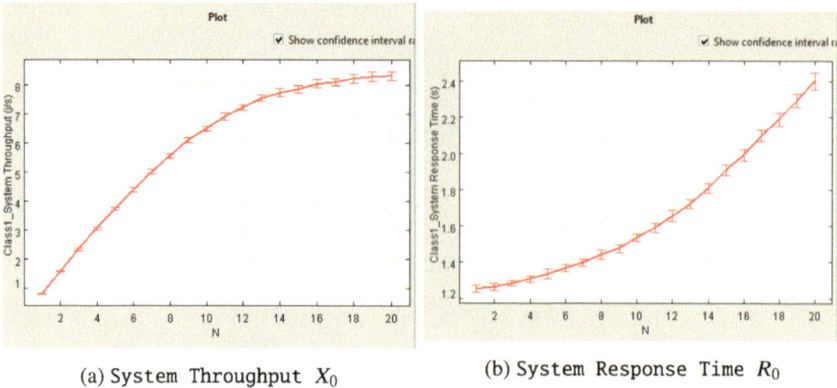

(a) System Throughput X_0 (b) System Response Time R_0

Fig. 2.9 System Throughput and System Response Time versus Number of customers

A What-if analysis is performed by setting the Number of customers $N_0 = 1 \div 20$ as *control parameter*. Figure 2.9 show the behavior of System Throughput X_0 and System Response Time R_0, respectively, with respect to N_0. Please note that the R_0 values computed by JSIMg *include* the time spent in the Reference station, i.e., the Users station, that is $Z = 1$ s. For $N_0 = 20$ we have $X_0 = 8.32$ req/s and $R_0 = 2.4$ s. As N_0 increases from 1 to 20, X_0 becomes flat and tends to its horizontal upper bound, while R_0 becomes linear and tends to its lower bound which is a oblique line. These behaviors are typical of closed systems when a resource is approaching *saturation*. In the following *Objs. 2, 3* we will analyze this condition in detail.

The values of some percentiles of the System Response Times, for example the 90th or the 95th, are often requested in performance studies. Let us recall that the 90th percentile Π_{90} of a variable Y is the value below which can be found 90% of all the values assumed by Y, i.e., it is $P(Y \leq \Pi_{90}) = 0.9$. To obtain the percentile values in JSIMg it is necessary to flag the check box Stat.Res. (see, e.g., Fig. 1.8) in the window of the metrics to be collected. A CSV file with *all* collected values of the selected metric is then generated and stored. Various statistical indexes are computed by clicking on the Statistical Results button (see Fig. 1.9) in the window of the analyzed metric. Selecting Distribution as a drawing option, the values are sorted in increasing order and are grouped in intervals. For example, 300 intervals have been selected in Fig. 2.10. The percentiles corresponding to each interval are calculated and stored in a CSV file. A sample of this file for the intervals $70 \div 76$ with the corresponding percentiles (from 88.9 to 91.3) is shown in Fig. 2.11. The 90.1 percentile corresponds to $R_0 = 4.88$ s. It should be noted that if the values of a variable Y are exponentially distributed it is $\Pi_{90} \simeq 2.3 \times$ (*mean value of Y*). In our case, the values of R_0 are hypo-exponentially distributed (the coefficient of variation is $0.76 < 1$, see Fig. 2.10). Their variance is less than that of an exponentially distributed variable with the same mean. Thus, it seems correct to obtain the value of

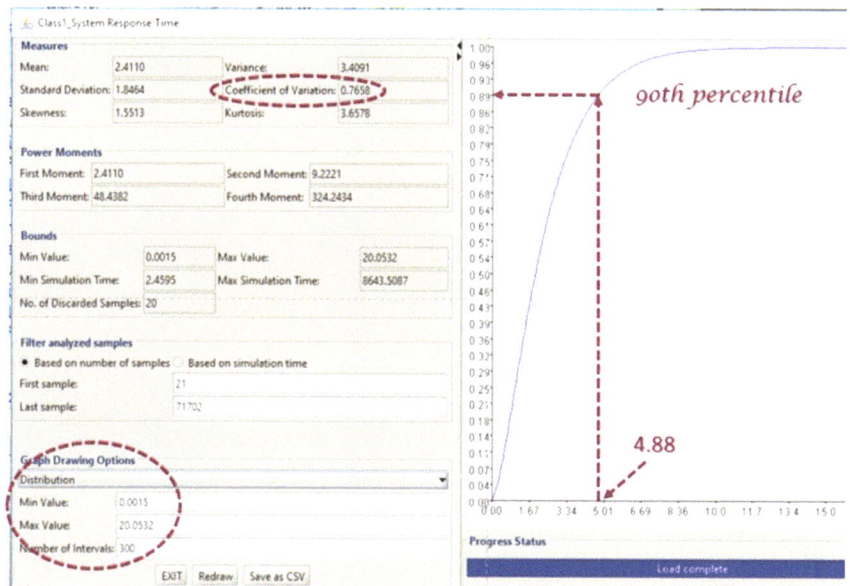

Fig. 2.10 Statistical indexes of the `System Response Times`

```
70  4.613391   4.68023  0.889174
71   4.68023  4.747069  0.893597
72  4.747069  4.813908  0.897754
73  4.813908  4.880747  0.901953  ◄----- 90.19th percentile
74  4.880747  4.947586  0.905887
75  4.947586  5.014425  0.909417
76  5.014425  5.081264  0.913198
```

Fig. 2.11 Sample of the CSV file with the values of R_0 sorted in increasing order and subdivided into 300 intervals. The four columns refer respectively to: the *id* of the intervals, the *extremes* of each interval, and the *percentile* corresponding to the extreme with maximum value

4.88 s for the 90th percentile of R_0 which is less than 5.52 s (2.3 x 2.4), as it would be if they were exponentially distributed. By increasing the number of intervals, more detailed percentiles can be obtained.

Obj.2: To improve the computing infrastructure performance, one of the first actions that seems natural is to replace AS_2, the slowest of the `App&Storage` servers, with a new model that is 20% faster (that is, the same as AS_1). Evaluate the effects on X_0 and R_0.

The mean `Service time` of server AS_2 of the original model must be modified decreasing its value from 0.025 to 0.020 s. The model with the `What-if` for $N_0 = 1 \div 20$ *users* is executed again.

As expected, the `Utilization` of AS_2 decreased, e.g., from 61.8% to 50% with $N_0 = 20$. However, surprisingly *NO improvements* are obtained on X_0 and R_0.

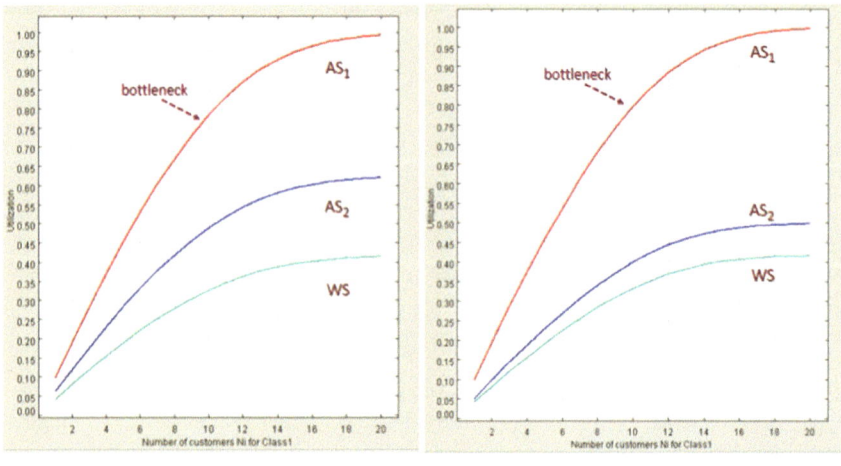

(a) Utilizations of the original servers   (b) Utilizations with the upgraded server AS_2

Fig. 2.12 Utilizations of the three servers AS1, AS2, and WS *versus* N_0

Indeed, with the new fast server AS_2 we have $X_0(20) = 8.27$ req/s and $R_0(20) = 2.42$ s while with the slow one we had $X_0 = 8.32$ req/s and $R_0 = 2.40$ s, respectively. The two values of X_0 can be considered equally likely estimates of the exact throughput value since they are both in the same 99% confidence interval. The same observation applies to R_0 (see Appendix A.2).

Analyzing the Utilizations of the three servers, in Fig. 2.12a with the original configuration and in Fig. 2.12b with the new AS_2, we have an answer to this *unexpected* result. From Fig. 2.12a it is possible to see that the utilizations of AS_1 and AS_2 are *unbalanced*, and that AS_1 is the *bottleneck* of the computing infrastructure despite being the *faster* of the two. Indeed, its utilization is the highest of all servers and for heavy load it is close to saturation (e.g., with $N_0 > 15$ it is $U_{AS1} > 0.95$).

This is the main motivation of the *uselessness* of the action we have done:

> *improving any station but the bottleneck do not generate any performance gain with heavy workload.* It is known that performance improvements can only be achieved by *reducing its contention.* Actions that reduce the load of stations other than the bottleneck produce *minimal improvements* (if any) only under very light workload (see *Obj.3*).

Obj.3: Given the insignificant results obtained in Obj.2, we want to evaluate the performance improvements that can be achieved by replacing the AS_1 server with a new model 20% faster (the same increase considered in Obj.2 for AS_2).
We recompute the original model (Fig. 2.7) settings the mean Service Time of server AS_1 to a value 20% faster (from 0.020 s to 0.016 s). We then execute again the What-if for $N_0 = 1 \div 20$ *users* obtaining the values of X_0 and R_0 reported in Fig. 2.13. For $N_0 = 20$, with respect to the original system, X_0 increases of 20%, from

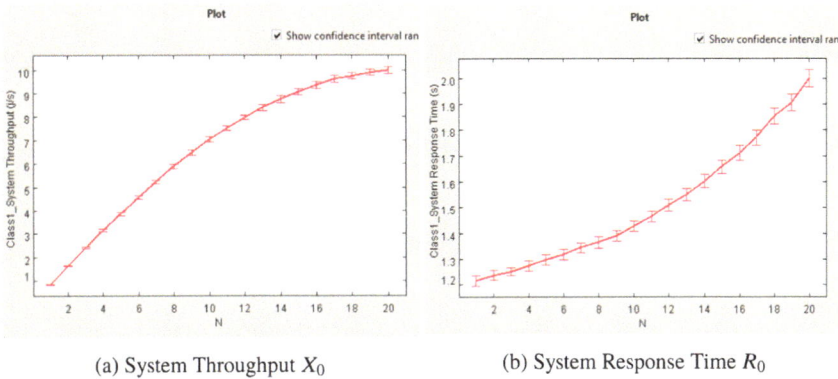

(a) System Throughput X_0 (b) System Response Time R_0

Fig. 2.13 X_0 and R_0 with the new server AS_1 20% faster

8.32 to 9.99 req/s, and R_0 drops of 17% from 2.4 to 1.99 s. The bottleneck remain the server AS_1, its utilization is 0.95 (in the original model was 0.99).

Let us remark that these *positive* results were obtained because we *improved* the station that is the *bottleneck* of the system, i.e., the server AS_1. Indeed, as seen in the previous *Obj.2*, improving other stations do not produce any significant results on performance.

Obj.4: According to the management, the number of internal employees authorized to access the computing infrastructure may increase to 40 in a semester. Which will be R_0 and X_0 with the actual configuration with $N_0 = 40$ users?

We recompute the original model (Fig. 2.7) settings the Number of Customers in the closed class definition window to 40. The behavior of the mean value of R_0 and of the confidence intervals during the simulation are shown in Fig. 2.14. As can be seen, the mean value of R_0 is 4.821 s obtained from the model is very close to the lower bound 4.8 s given by $N_0 D_{max} = N_0 V_{AS1} S_{AS1} = 40 \times 0.12$. The $X_0(40)$ is 8.325 req/s, very close to its upper bound $1/D_{AS1} = 1/0.12 = 8.333$ req/s (see Sect. 2.3).

2.3 Equivalent Model with Service Demands

tags: closed, single class, Delay/Queue, JSIMg

In this section we describe a model, solved with JSIMg, parameterized with Service demands. This model is *equivalent* to the one solved in the previous section using the Visits and Service times of the stations. The granularity at the *system-level* is here adopted compared to that at the *station-level* adopted in the previous model.

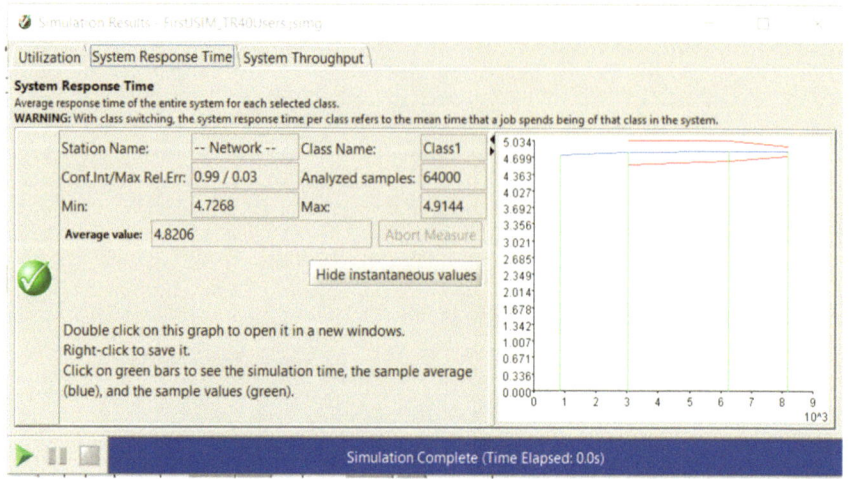

Fig. 2.14 `System Response Time` of the computing infrastructure with the initial configuration and $N_0 = 40$ users

2.3.1 Problem Description

The models can be designed at different levels of granularity (see Chap. 1), from the single component (station) to the entire system. In Sect. 2.2 we implemented a station-level model using the `Routing probabilities` p_{ij}, i.e., the probability that a request in output from station i is routed to station j. We have also described how to obtain from the p_{ij}s the number of `Visits` V_r that a request makes to each station during its complete execution (see Appendix A.1). In the models parameterized at this level of detail we may compute all the performance indexes describing the behavior of each station, including its `Throughput` and `Response time`. However, measuring the p_{ij}, or the V_r, is difficult or in some cases impossible. A parameter that is often used is the `Service demand` D_r of a request to station r, which represents the *total amount* of `Service time` that a request requires from station r to complete its execution. The D_r values may be obtained by multiplying the `Service time` S_r required by one visit to the number of `Visits` V_r that a request makes to station r during its execution, i.e., $D_r = S_r V_r$. There are many motivations that make the models parameterized with `Service demands` so *popular*. Among them are:

- the *limited effort* required to obtain the mean values of `Service demands` D_r from *measurements*. Indeed, the system log file usually shows the D_r values for every request executed. Recall that several executions are needed to have a *reliable estimate* of the mean values of D_r and the *confidence intervals* of the measured values must be computed, (see Appendix A.2 and, e.g., [36, 37]). The D_rs can also be obtained dividing the *busy time* B_r of a resource by the number C_0 of user requests executed;

- the models that use Service demands are less expensive to parameterize than more detailed models using Service times and Visits as the number of parameters required is significantly lower. The measurement of even one more parameter may require a non-trivial effort;

- a large part of queueing networks considered in performance studies are of the *separable* type (see Sect.1.2 and, e.g., [4, 25]) and can be solved by knowing *only* the values of D_r and not those of its single factors V_r and S_r. The paths followed by requests between the resources can be unknown, only the *global amount* of service time required to each resource (i.e., the Service demands) counts. According to this property, for example, a model in which a job make 1000 visits to a station whose service time is 5 milliseconds, is equivalent to one in which the job make a single visit to that station requiring 5 sec of service time. Clearly the equivalence must be applied also to all the other stations of the system. The performance indexes obtained with this equivalent model are the same as the more detailed model with regard to the indexes at the system-level, i.e., System Throughput and System Response Time. The same is true also for the Utilization and the Residence time of the single stations. However, in this case, due to the *high level* of granularity adopted, we *cannot compute* the Throughput and the Response time at the station-level (their computation requires the Visits and Service times of each station).

Based on the described advantages, when possible, the models are preferably parameterized in terms of Service demands D_r instead of Visits and Service times. Clearly, with this high level of granularity we lose the structural similarity with the considered system, but the models are easier to implement, the solution algorithms are faster, and the performance indexes that can be computed (not all, but almost) are correct.

2.3.2 Model Implementation

To illustrate the practical applicability of the Service demands we consider again the closed model solved in Sect.2.2 using the Routing probabilities p_{ij} and the Service times S_r, see Fig. 2.6b. In this Section we implement a new version of it using Service demands. From the p_{ij}s it is possible to derive the Visits to the three servers $V_{WS} = 10$, $V_{AS1} = 6$, $V_{AS2} = 3$ (see Appendix A.1) and knowing the Service times $S_{WS} = 0.005$ s, $S_{AS1} = 0.020$ s and $S_{AS2} = 0.025$ s, we can compute the Service demands ($D_r = V_r \, S_r$) $D_{WS} = 0.050$ s, $D_{AS1} = 0.120$ s, $D_{AS2} = 0.075$ s.

The implemented model of Fig. 2.15 consists of three servers, having as Service times the D_r, which are visited only once during the execution of a user request. The structure of this new model is clearly simpler than that of Fig. 2.7.

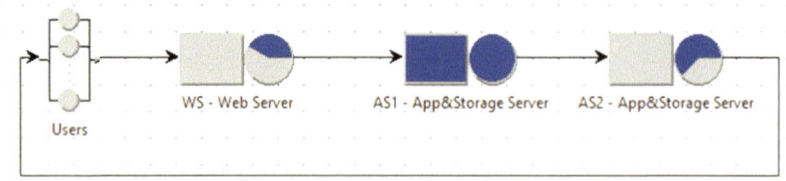

Fig. 2.15 Model parameterized with `Service` demands equivalent to that of Fig. 2.7 which uses `Visits` and `Service times`

2.3.3 Results

We solve the model of Fig. 2.15 using the `What-if` analysis with $N_0 = 1 \div 20$ `customers` as `control parameter`. The performance measures obtained are practically the same, i.e., they lie in the same 99% confidence interval of the corresponding indexes obtained with the model of Fig. 2.7. Table 2.1 compares the values of some performance indexes computed by the two models with $N_0 = 20$ *customers*.

It should be noted that with the high level parameterization, and the consequent simplified layout of the equivalent model, it is NOT possible to compute the `Throughput` and the `Response time` of *each station*. Indeed, at the system level, we do not model the `Visits`, thus *only* the parameters at high level of aggregation can be computed for each station, i.e., the global `Utilization` and the `Residence time`. The values inside the boxes in Table 2.1 emphasize that in the equivalent model the three servers have the *same* `Throughput` (measured in user requests per time unit), that coincide with the `System Throughput` X_0. This is because the three servers in this model are visited only once, requiring the `Service` demand D_r to each of them, are connected in series. Thus, we may compute only the `Residence times` Rd_r of the servers and *not* their `Response times` R_r (since it is $Rd_r = V_r R_r$).

Table 2.1 Performance metrics for $N_0 = 20$ *customers* obtained from the two equivalent models parameterized at different levels of granularity (`Visits and Service times` and `Service demands`, respectively)

Parameters used	Performance metrics										
	R_0	X_0	U_{WS}	U_{AS1}	U_{AS2}	Rd_{WS}	Rd_{AS1}	Rd_{AS2}	X_{WS}	X_{AS1}	X_{AS2}
Visits and Service time (Fig. 2.7)	2.4	8.32	0.419	0.993	0.618	0.082	1.14	0.184	82.13	49.32	24.74
Service demands (Fig. 2.15)	2.4	8.31	0.415	0.992	0.622	0.083	1.14	0.185	8.31	8.31	8.31

The *operational analysis laws* can be applied also to models parameterized with Service demands. For example, using the *Forced Flow law* $X_r = X_0 \, V_r$, the *Utilization law* becomes

$$U_r = X_r \, S_r = X_0 \, V_r \, S_r = X_0 \, D_r \qquad (2.3)$$

From Table 2.1 it is possible to see that Eq. 2.3 is verified with the metrics obtained from the model parameterized with the D_r. For example, for the server WS it is $U_{WS} = 8.31 \times 0.05 = 0.415$ that coincides with the measured value of U_{WS}. Note that the results of the two models may not coincide exactly as we are in simulation and we know only the confidence intervals of the computed variables. We may verify also that, according to Eq. 2.3, the ratio of the U_r coincides with the ratio of the D_r:

$$\frac{U_{WS}}{U_{AS1}} = 0.418 \simeq \frac{D_{WS}}{D_{AS1}} = 0.416 \qquad \frac{U_{AS1}}{U_{AS2}} = 1.594 \simeq \frac{D_{AS1}}{D_{AS2}} = 1.6 \qquad (2.4)$$

The *Little law* applied to resource r using the Residence times becomes:

$$N_r = X_r \, R_r = X_0 \, V_r \, R_r = X_0 \, Rd_r \qquad (2.5)$$

It should be recalled that the System Response Time R_0 provided by JSIMg comprises the time spent by a user request in the Reference station, that in our model is the Users with $Z = 1$ s. So, it is $R_0 = Rd_{WS} + Rd_{AS1} + Rd_{AS2} + Z = 2.409$ s. Applying *Little law* at the system-level we have: $N_0 = X_0 \, R_0 = 8.31 \times 2.409 = 20 \, customers$, as expected.

From the analysis of the D_r we can derive that the server AS1 is the *most utilized* of the resources since $D_{AS1} = 0.120$ s is the *largest* of the Service demands. As N_0 increases it will be the first resource to saturate, i.e., it becomes the *bottleneck* of the system, limiting the System Throughput to $X_0 \leq 1/D_{max} = 8.333$ req/s. With $N_0 = 20 \, customers$ we obtained $X_0 = 8.31$ req/s (see Table 2.1) since AS1 is not completely saturated (it is $U_{AS1} = 0.992$).

2.4 Optimal Operating Point of a Server

tags: open, single class, Source/Queue/Sink, JSIMg.

We describe how to identify the optimal operating condition of a system that is characterized by the highest Throughput with the shortest Response time. A system in this condition, referred to as *optimal operating point*, operates with *maximum of efficiency*, that is, it maximizes its productivity by introducing the minimum delay. We consider a simple model of a system that is solved using both analytical techniques (Queueing Networks) and simulation techniques (JSIMg model).

2.4.1 Problem Description

Identifying the *optimal operating point* of a digital infrastructure is a problem that IT managers face in their daily lives. Nowadays, this task is becoming more and more important as the size of data centers and clouds continues to grow in the number of servers, applications and users. The basic idea is that a small profit on a single server can translate into a large profit when evaluated on hundreds of servers with thousands of users.

Clearly, depending on the context considered, the notion of *optimal operating point* assumes various definitions that translate into different actions. For example, it can refer to the operating conditions that minimize the energy required to execute a workload while meeting the performance goals, or to the operating point that satisfies the SLA (Service Level Agreement) by minimizing the number of allocated servers.

To simplify the description, let's consider a single server that we assume is operating under the *optimal* conditions when its Throughput X is *maximized and* its Response time R is *minimized.* The load corresponding to this *optimal condition* will be referred to as *optimal load.*

In this simple case, when the goal of the performance study is to identify the load that *maximizes* the Throughput X or the one that *minimizes* the Response time R, the answers are easily provided. In fact, a resource generates the *maximum* X when it is *saturated* and provides the *minimum* R when only one request is in execution, that is, there is *no contention.* However, when X and R are to be compared at the *same time*, a new metric must be used that considers the trade-off between the two. To this end, below we will consider the System power Φ, a metric introduced in [19] and extensively studied by Kleinrock [23, 24], defined as

$$\Phi = \frac{X}{R} \tag{2.6}$$

The behavior of Φ may be considered in some way related to that of the *Quality of Service.* Indeed, an increase in Throughput or a decrease in Response time increases the System power, that may be considered in the *SLA* as indicator of the *Quality of Service* delivered to the users.

In the next section we consider a system consisting of a single server modeled with a Queue station (see, e.g., Fig. 2.1) that execute a homogeneous workload (single class) with Interarrival times and Service times exponentially distributed.

For this simple case, we describe the analytical computation of the optimal load, and then we implement the correspondent model with JSIMg. The analytical derivation of Φ for more complex systems is not easy to obtain. However, it should be emphasized that the JSIMg provides the System Power behavior for *all simulated models*, regardless of *their complexity.* Φ is one of the metrics available in the tool.

2.4.2 Model Implementation

In this section we address the problem of the identification of the optimal operating point of a single server executing requests with `Interarrival times` and `Service times` exponentially distributed. For its simplicity, we initially present the analytical solution of this model that was derived in [23]. Below, we describe the simulation results obtained with the corresponding model implemented with JSIMg.

The `System Power` Φ for the considered M/M/1 model is

$$\Phi = \frac{\lambda}{R} = \frac{\lambda\,(1 - \lambda S)}{S} \tag{2.7}$$

where λ is the arrival rate and S is the mean service time of the requests. To find the load λ^{opt} that *maximizes* Φ it is sufficient to set to zero its first derivative Φ' with respect to λ and derive the value of λ^{opt}. We have: $\Phi' = (1/S) - 2\lambda^{opt} = 0$, thus it will be

$$\lambda^{opt} = \frac{1}{2}\frac{1}{S} \tag{2.8}$$

Therefore, according to Eq. 2.8, the *optimal operating point* is obtained with a load λ^{opt} equal to *half* of the one corresponding to the `maximum Throughput` 1/S. The R^{opt}, U^{opt} and N^{opt} are

$$R^{opt} = \frac{S}{1 - \lambda^{opt}S} = 2S \qquad U^{opt} = \lambda^{opt}S = 0.5 \qquad N^{opt} = \frac{0.5}{1 - 0.5} = 1\ req$$

Let us remark that R^{opt} is *twice* its minimum value S, the server is utilized at 50% and the mean number of customers in the server is 1, 0.5 in queue and 0.5 in execution. In this optimal condition, an arriving request has 50% of probability to find the queue empty and the server idle. Figure 2.16 shows the behavior of Φ of a server with $S = 1$ s.

An interesting observation can be obtained from the analysis of Fig. 2.16b which shows that `System Power` is maximized with the load corresponding to the *tangent point* of the straight line $R = m\lambda$ from the origin to the `Response time` curve. Equating to zero the discriminant of the equation that compute the intersection of the two functions we obtain $m = 4S^2$. Replacing it in the equation of the intersection we obtain $\lambda = 0.5(1/S)$, which has already been found as *optimal load* in Eq. 2.8.

This property allows to define the optimality condition as the one corresponding to a load λ^{opt} for which the relative increase of the `Throughput` X is equal to that of the `Response time` R. When it is $\lambda < \lambda^{opt}$ it will be $dX/X > dR/R$ therefore an increase in λ increases the `Throughput` more than the `Response time`, so the gain is higher than the loss. The opposite situation occurs when it is $\lambda > \lambda^{opt}$ since in this condition an increase of λ generates losses greater than the gains, i.e., it is $dX/X < dR/R$.

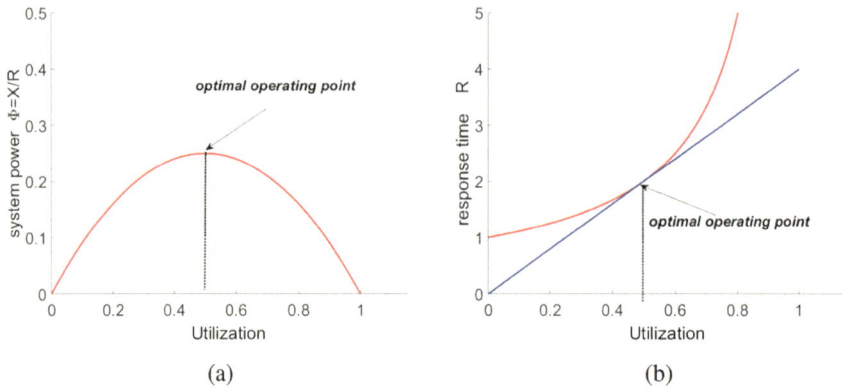

Fig. 2.16 Power Φ (**a**) and Response time R (**b**) versus Utilization with S = 1 s

This definition of the optimal operating point is valid for *any* Throughput and *any* Response time functions. The metric System Power can be very useful to implement load balancing policies based on machine learning and as a target function in autoscaling components.

The implemented JSIMg open model consists of three stations: Source, Queue, Sink (see, e.g., Fig. 2.1). The Service times of the Queue are exponentially distributed with mean $S = 1$ s. The Interarrival times of the requests generated by the Source have exponential distribution with arrival rates ranging from 0.1 to 0.9 req/s. Figure 2.17 emphasizes the selection of System Throughput, System Response time, and System Power indexes (the last two are shown in the graphs of Fig. 2.18).

2.4.3 Results

A What-if analysis is used with the arrival rate λ of requests as *control parameter* with values ranging from 0.1 to 0.9 in 9 models. As can be seen from Fig. 2.18, the values of R and Φ corresponding to the optimal load $\lambda^{opt} = 0.5$ req/s are very close to the exact ones obtained analytically ($R = 2$ s, and $\Phi = 0.25$). The confidence intervals are very small.

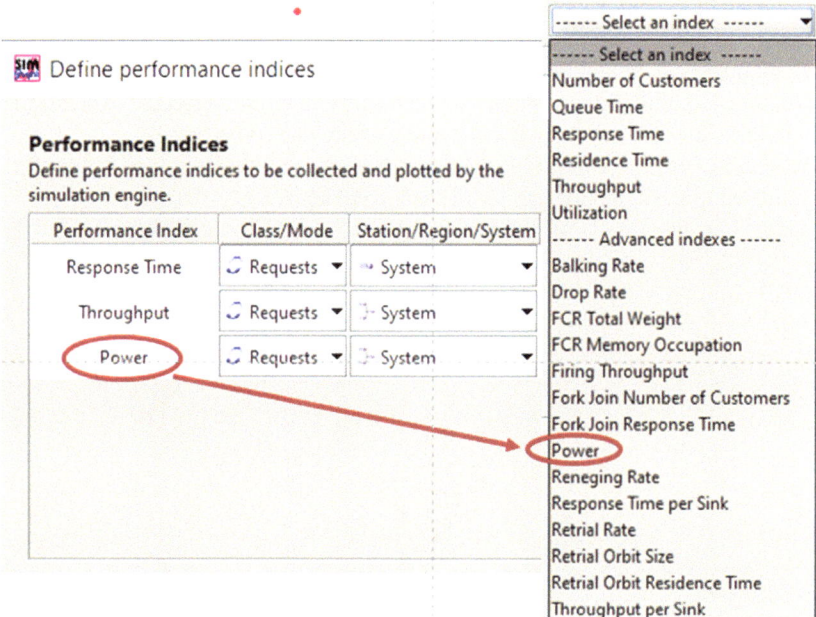

Fig. 2.17 Selection of the `System Power` index

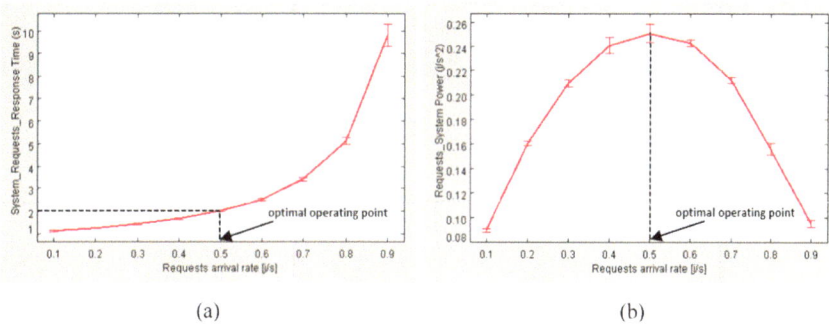

(a)   (b)

Fig. 2.18 `Response time` R and `System Power` Φ obtained with JSIMg

2.4.4 Limitations and Improvements

- *High variability of* `Service times`: In [24] it is described the analytical derivation of Φ in simple models with exponential `Interarrival times` and high
  variability of `Service times` (for M/G/1 stations).

- *Models with complex structure*: Although we have described the use of `System power` in models with a single resource, it should be clear that all the considerations made can be applied also to open and closed models with *more complex* structure and multiple resources. The identification of the analytical expression of Φ in these models is clearly not so simple as the one of Eq. 2.7.
- *Availability of Power index*: JSIMg compute and plot the values of Φ for all the simulated models, *independently* of their complexity.

Chapter 3
Systems with Heterogeneous Workloads

3.1 Parameterization of Heterogeneous Workloads

As for the *characterization of the requests*, there is a direct correspondence between Service times, Visits, and Service demands used in *single-class* models and those used in *multiclass* models. However, since their values must be specified on a *per-class* basis, each parameter must be identified now with *two* indexes: the station and the class it will refers to. For example, the service demand D_r of resource r becomes now $D_{r,c}$, the Service demand of class-c request to resource r.

However, new problems arise when the growth of workload intensity has to be described as this can be done in different ways. Indeed, while to specify the workload intensity in single-class models is *sufficient* to know the Number of customers N_0 in execution in closed systems, or the arrival rate λ_0 and distribution of Interarrival times in open systems, the presence of multiple classes make these descriptions *no longer adequate*. Recall that with the index 0 (*zero*) of a metric we refer to the system as a *whole*.

In this section we first consider *closed* models, then we will analyze *open* models.

In closed models with of C *classes* of jobs, the workload intensity is described by the vector $\mathbf{N_0} = \{N_{0,1}, N_{0,2}, ..., N_{0,C}\}$ whose components are the number of jobs of each class in execution. The total number of jobs in execution is given by $N_0 = N_{0,1} + N_{0,2} + ... + N_{0,C}$. For example, $\mathbf{N_0} = \{25, 75\}$ means that in the closed model there are globally $N_0 = 100$ *jobs* in execution, 25 of class-1 and 75 of class-2.

A *new* parameter very useful for the description of multiclass workloads *growth* is the vector $\boldsymbol{\beta}$ representing the *fractions of jobs* of the C classes in execution in the system, that we will denote as *population mix* or *job-mix*:

$$\boldsymbol{\beta} = \{\beta_1, ..., \beta_C\} \quad \text{with} \quad \beta_c = N_{0,c}/N_0 \quad \text{and} \quad \beta_1 + \beta_2 + ... + \beta_C = 1 \quad (3.1)$$

Using the population mix, the workload $\mathbf{N_0}$ of the previous example can be described by $\mathbf{N} = N_0\,\boldsymbol{\beta}$, with $N_0 = 100$ and $\boldsymbol{\beta} = \{0.25, 0.75\}$.

© The Author(s) 2024
G. Serazzi, *Performance Engineering*,
https://doi.org/10.1007/978-3-031-36763-2_3

The importance of the *job-mix* lies in the fact that in multiclass networks the ratio of the global utilizations of two stations is no longer *constant* with N_0, as in single class case, but depends on the fractions of jobs of the various classes in execution. Indeed, applying in *single-class* models the *Utilization law* to resources i and j we have: $U_i = X_0 D_i$ and $U_j = X_0 D_j$, and their ratio $U_i/U_j = D_i/D_j$ is *constant* with N. The immediate consequence of this behavior is that the *bottleneck* of the system may *migrate* among the resources as a function of the population mix. Thus, since it is known that the overall performance of a system is limited by the congested resource (i.e., the bottleneck), the fluctuation of the mixes may abruptly change them deeply.

While the definition of bottleneck is simple, i.e., the resource with the highest utilization, in multiclass models, the problem of bottleneck identification is not trivial since the same model can exhibit different bottlenecks depending on the population mix. Different types of bottlenecks can be identified. The *class-c bottleneck* is the station with the highest service demand of that class and saturates (its utilization tends to one) when the number of class-c customers grows to infinity. The problem in multiclass systems is that, as a function of the population mix, a station may saturate also if it is **not** a *class-bottleneck* (in this case the station will be referred to as *system-bottleneck* or *model-bottleneck*) or **more** stations may saturate *concurrently* with several mixes, referred to as *common saturation sector* (see [2, 3, 15]).

Therefore, to characterize the workload behavior in multiclass models, we must describe the variations of the mixes. In general, *different* β may yield *different* bottlenecks. Two types of workload increment should be considered: *proportional* and *unbalanced*. The population growth that consists of letting the total number of customers N_0 to grow keeping *constant* the population mix β is referred to as *proportional growth*.

According to this type of growth, in the example of workload above considered, the jobs in execution will be increased according to the proportions 25% of class-1 and 75% of class-2 since the mix is $\beta = \{0.25, 0.75\}$. So, when the total number of jobs increases to 300, we will have 75 class-1 and 225 class-2 jobs in execution.

We have the *unbalanced population growth* when only one class of jobs, say c, increases. As $N_{0,c}$ continue to growth, the bottleneck of the system tends to the station that is the *class-c bottleneck*, the population mix tends to $\beta = \{0, 0, 0, ..., 1_C\}$, and the performance tend to the asymptotes of single-class workloads.

To support users who need to model the different types of population growth, JMT implement specific features of the `What-if` analysis that allow the automatic increment of the population of a single class only (see, e.g., Fig. 3.2) or the generation of all the possible mixes of two classes in closed models (see, e.g., Fig. 3.7b).

Most of what has been previously described for the closed models also applies to *open models*. The number of jobs in execution of the various classes must be replaced with the corresponding arrival rates. So, the global arrival rate to a open system is $\lambda_0 = \{\lambda_{0,1}, \lambda_{0,2}, ..., \lambda_{0,C}\}$ and the population mix is described by:

$$\beta = \{\beta_1, ..., \beta_C\} \qquad \beta_c = \lambda_{0,c}/\lambda_0 \qquad \beta_1 + \beta_2 + ... + \beta_C = 1 \qquad (3.2)$$

Differences between open and closed systems lie in the *bottleneck switch* behavior. In the former, the bottleneck migrate *instantaneously* between two resources without going through a *common saturation section*, i.e., the set of mixes that saturate both concurrently. Regarding the *scheduling disciplines* of the various classes in multiclass models, there are differences as a function of the solution technique adopted. While with *simulation* the users have practically *no limitations* (some minor incompatibilities may take place as a function of the types of discipline selected), the analytical technique introduce some constraints. Typically, the queueing networks that are solved analytically are of the *separable* types (see, e.g., [9, 36]) and their solution (the stationary probability of their states) can be obtained by the product of the individual solutions of the stations. The computational complexity introduced by the presence of multiple classes that may have different scheduling algorithms, and usually by the large numbers of stations and customers has required the introduction of some limitations. An important theorem, the BCMP (see, e.g., [6, 36]), for *open*, *closed*, and *mixed* queueing networks define the characteristics that a multiclass network should have to be separable and thus to be solved analytically with efficient algorithms. Each station must be of one of the following types:

- a *queue* station with *FCFS scheduling discipline* (requests are served according to the sequence of arrival), with one or more servers, having the *same* exponential distribution of Service times for all the classes. For each resource, the Service times $S_{r,c}$ must be the *same* for *all* the classes. The differences between the classes may be considered using the number of *visits* $V_{r,c}$ to the resources, providing the possibility to have different Service demands $D_{r,c}$ for the same station (which in any case *cannot* be modeled with FCFS discipline).
- a *queue* station with PS (*processor-sharing*) scheduling discipline: the n requests in the station are served simultaneously receiving each $1/n$ of the server capacity. For example, in a queue station with one server if during the execution of a request that has Service time S = 2 s there are 10 requests in the station, then the execution of this request will be completed after 20 sec. This discipline is commonly used to model the time quantum of the processors, in this case the quantum tends to zero. The distribution of the service times, and their *means*, can be *general* and *different* for *each class*.
- a *queue* station with LCFS-PR (*last-come first-served preemptive-resume*) scheduling discipline. When a new request arrives, it interrupts the execution currently in progress and starts its execution immediately. When it is completed, the last preempted request resumes the execution at the point it was interrupted. The distribution of the Service times, and their *means*, can be *general* and *different* for *each class*.
- a *delay* station, referred to as IS *infinite servers* station, in which each request has its dedicated server. Its Response time coincide with the Service time since there is no queue time. The distribution of the service times, and their *means*, can be *general* and *different* for *each class*.

Load-dependent service times are allowed. For PS, LCFS-PR, and IS stations the service times for the requests of a class may depend on the number of requests of that class in the station, or on the global number of requests in that station. For FCFS station the service times may depend only on the global number of requests of all the classes in that station.

In conclusion, users *must be sure* that the multiclass models who want to solve *analytically* have stations of the four types described. For example, if you want to solve analitycally a model that has different per-class `Service times` $S_{r,c}$ on the same resource, the JMVA will solve it in any case (it compute the $D_{r,c}$) but you should be aware that the modeled scheduling discipline is PS and not FCFS. If in any case it is necessary to solve this model with the FCFS discipline, then there is no other choice than to use simulation.

3.2 Motivating Example of Multiclass Models

tags: closed, two-class, Delay/Queue, JMVA.

In this section we describe an example *purposely designed* to emphasize the errors on the performance forecast that can be obtained from the *same model* assuming that a multiclass workload consists of a single class of jobs. The counterintuitive behavior exhibited by some performance indexes of multiclass models as a function of the fluctuations of the classes of jobs in execution is also investigated. We solve this model analytically with JMVA.

3.2.1 Problem Description

Consider a powerful web server, accessed by administrative staff and graduate students of a university, which has two main resources: *CPU* and *Storage* (Fig. 3.1a). The workload consists of *two* different applications. The first one is used to manage the *administrative* procedures concerning the students curricula (tuition fees payments, courses attended, grades obtained, ...). The second one is devoted to the management (uploads, downloads, folder structure) of the course materials/*documents* (slides, notes, class exercises, homeworks, exams) that professors, assistants, and students access. According to the resource requirements of the two types of users, *two* classes of jobs, referred to as *Adm* and *Doc*, are identified.

Initially, we want to investigate the effects on performance indexes generated by different workload scenarios. More precisely, we increment the number of jobs of *one class only* while keeping constant the jobs of the other class. The different growth rates of the various classes of the workload are described by modifying the parameter β. To model the growth of class-i jobs, we increase the corresponding β_i.

Then, we want to illustrate the errors that can be introduced in the performance forecast by a wrong assumption on the number of classes of the workload. We solve the same system model assuming the two-class workload as consisting of a

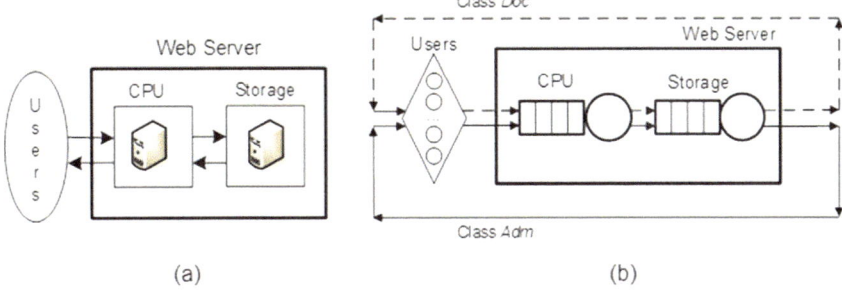

Fig. 3.1 Web server layout (**a**) and closed queueing network with two different applications (**b**)

single-class of jobs. We consider a `base` system and an `upgraded` system with a CPU more powerful of a factor of five (the corresponding CPU service demands are decreased by five times). The behavior of the performance indexes are studied with respect to all the possible combinations of the two classes of jobs in execution, i.e., all the possible population mix.

3.2.2 Model Implementation

We use a *closed* model (Fig. 3.1b) since the customers that have access to the server are limited (administrative staff and graduate students). It consists of three resources: `Users`, `CPU`, and `Storage`. The `Users` station is of *delay* type.

The two classes of the workload are characterized by the `Service demands` shown in Table 3.1. Let N_0 be the global number of jobs of the two classes. To model the growth of class-*Doc* jobs only (*unbalanced* population growth), we use the `What-if` feature with `Number of customers` and class-*Doc* as `control parameters` (Fig. 3.2). The $N_{0,Doc}$ values range from 5 to 280 with step of 5. To study the effects on performance forecast corresponding to the two different assumptions on the number of classes (*one* and *two*) in the workload, we consider two configurations, i.e., `base` and `upgraded`, of the same system. The behavior of performance indexes in the two configurations has been investigated modeling

Table 3.1 `Service demands` [s] of the *two classes* of jobs

Resources (stations)	Two classes	
	Adm	*Doc*
`Users think time`	3	10
`CPU`	0.20	0.100
`Storage`	0.050	0.60

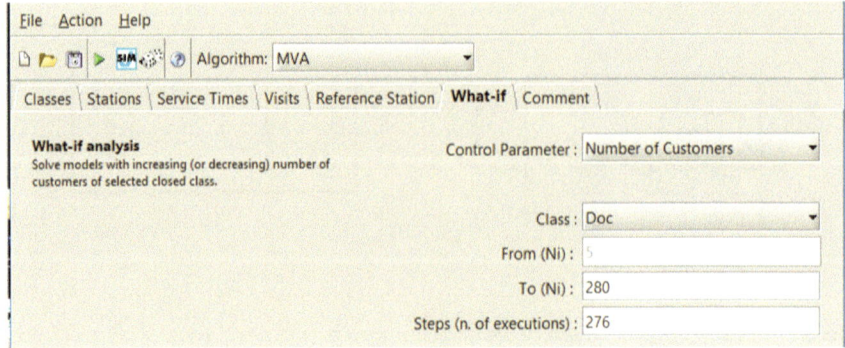

Fig. 3.2 Parameters of the `What-if` for the description of the *unbalanced* population growth: only class-*Doc* jobs increase while class-*Adm* jobs are kept constant

all possible population mixes $\boldsymbol{\beta}$ with $N_0 = 300$ jobs (Fig. 3.5) and modifying the `Service` demands of Table 3.1 (see Table 3.2).

3.2.3 Results

In what follows we will describe the operations required to achieve the *objectives* of the study (referred to as *Obj.1–Obj.2*).

Obj.1: Show the counterintuitive result that with a multiclass workload the Global System Throughput X_0 can decrease in spite that the global number N_0 of jobs in execution increases

We consider the model with the two-class workload whose service demands are shown in Table 3.1. The initial workload is $\mathbf{N} = \{20, 5\}$, globally $N_0 = 25$ jobs are in execution, 20 of class-*Adm* and 5 of class-*Doc*. The volume of traffic due to the class-*Doc* jobs is expected to increase during the next semester up to 280. This behavior is the typical *unbalanced population growth* that can be used when one class increases more than the others. We use the `What-if` feature of JMVA to evaluate the `Global System Throughput` X_0. The parameters that describe the increase of class *Doc* jobs are shown in Fig. 3.2. The `Control Parameter` is the `Number of customers` of class-Doc $N_{0,Doc}$, and the execution of 276 models with its increasing values from 5 to 280 are required.

In Fig. 3.3a the behavior of the `Global` and `per-class System Throughput` are shown for the number of jobs N_0 in execution increasing from 25, $\mathbf{N} = \{20, 5\}$, to 300, $\mathbf{N} = \{20, 280\}$. Initially X_0 increases until the number of *Doc* jobs reaches 19, corresponding to the maximum value of $X_0 = 5.416$ j/s. Then, any further increase of $N_{0,Doc}$, and clearly of N_0, corresponds to a decrease of X_0 (with $N_{0,Doc} = 280$ it is $X_0 = 2.724$ j/s). *How is it possible this happens?*

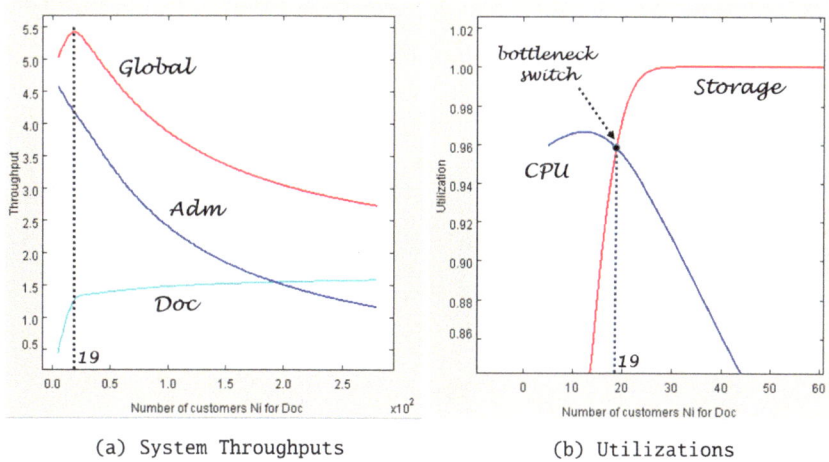

(a) System Throughputs   (b) Utilizations

Fig. 3.3 *Counterintuitive* behavior of performance indexes (Base system) with *unbalanced* increase of the Global Number N_0 of jobs in execution: only class-*Doc* jobs increases from 5 to 280

The answer is prompted by Fig. 3.3b showing the Global Utilization of CPU and Storage. For $N_{0,Doc} \leq 18$ the CPU is the most utilized resource, while for $N_{0,Doc} \geq 19$ the Storage is the most utilized. We are addressing the *bottleneck switch* phenomenon that can occur with multiclass workloads when different classes have their highest service demands on different stations. The basic concept is the following: the service demands of the various classes at the bottleneck station determine the performance of the global system. Since when the station bottleneck changes typically also the corresponding service demands are different, this migration may have a deep impact on the performance. While the *identification of the bottleneck* in single-class models is easy, in multiclass models is more complex [3]. As described in Sect. 3.1, with multiclass workloads the bottleneck may migrate among stations as a function of the percentage of jobs of the different classes in execution, i.e., of the *population mix*.

With the workload behavior considered in this *Obj.1* study, only class-*Doc* jobs increase from 5 to 280 and the population mix range from $\beta = \{0.8, 0.2\}$ ($\mathbf{N} = \{20, 5\}$) to $\beta = \{0.066, 0.933\}$ ($\mathbf{N} = \{20, 280\}$). When the *Doc* jobs are ≤ 18, the contribute of the *Adm* jobs, with $D_{max,Adm} = 0.2$ on CPU, is fundamental for the saturation of CPU. When *Doc* jobs, with $D_{max,Doc} = 0.6$, are ≥ 19 the load generated on Storage makes its global utilization predominant with that of the CPU.

As the number of *Doc* jobs continue to increase, asymptotically it will be ($N_{0,Doc} \rightarrow \infty$), and the workload assume the characteristics of a *single-class* with $\beta \rightarrow \{0, 1\}$. In this case, the maximum system throughput is given by $1/D_{max,Doc}$. In our workload the max Service demand of class-*Doc* is $D_{Sto,Doc} = 0.6$ s, thus it will be $X_0^\infty = 1/D_{Sto,Doc} = 1.666$ j/s. As can be seen from Figs. 3.3a, 3.4a, the System throughput of class-*Doc* and of the Global system tend to this

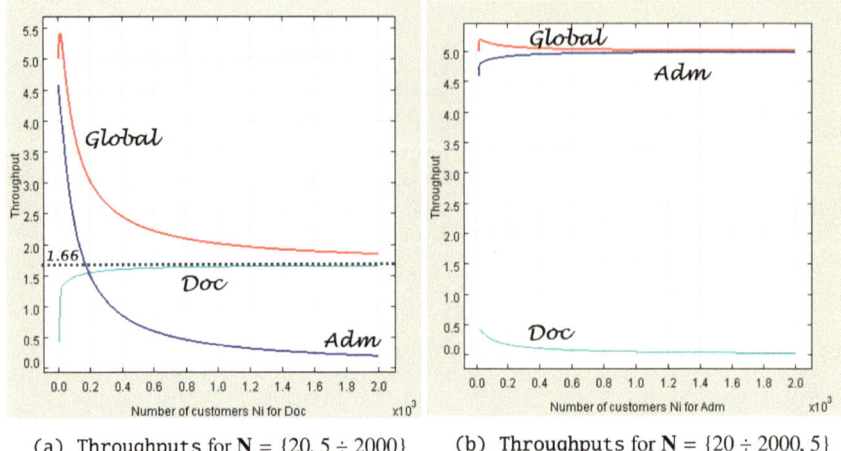

(a) **Throughputs for N** = {20, 5 ÷ 2000}   (b) **Throughputs for N** = {20 ÷ 2000, 5}

Fig. 3.4 System throughput asymptotes of the (Base system) when only class-*Doc* jobs increase from 5 to 2000 (**a**), and only class-*Adm* jobs increase from 20 to 2000 (**b**)

value: for $N_{0,Doc} = 280$ *jobs* it is $X_{0,Doc} = 1.57$ job/s and for $N_{0,Doc} = 2000$ *jobs* it is $X_{0,Doc} = 1.65$ job/s. The Global System Throughput X_0 decreases as $N_{0,Doc}$ increases beyond 19 jobs.

To emphasize the impact of the population mix on the Global System Throughput we ran two experiments with unbalanced population growth. Starting from the same initial workload **N** = {20, 5}, in the first one we increase to 2000 only class-Doc jobs while in the second one we increase to 2000 only class-Adm jobs. The System throughputs are shown in Fig. 3.4.

In Fig. 3.4a $X_{0,Doc}$ and X_0 tend to the same asymptotic value 1.666 *j/s* while in Fig. 3.4b $X_{0,Adm}$ and X_0 tend to the same asymptotic value 5 *job/sec*. The Global System Throughput X_0 tend to $X_{0,Doc}$ in (a) and to $X_{0,Adm}$ in (b). The differences between the two asymptotic values are evident. It should be pointed out that these values are *not* bounds! Indeed, a program mix that maximize the Utilization of *all* the resources of a system (see Fig. 3.3) maximize also the System throughput.

Obj.2: Show that assuming a multiclass workload as single-class allows the construction of models that generate very inaccurate performance forecast. Some counterintuitive results (other than those of Obj.1) that occur with multiclass models are also shown.

In this study (inspired, with some differences, by [25]) we will show that the performance projections obtained using a *wrong assumption* for the workload characterization, i.e., the workload is assumed to consist of a single class instead of multiple classes of customers, are *unreliable*.

We consider the closed system with three stations (Fig. 3.1b), that process the two-class workload whose Service demands are shown in Table 3.1.

Table 3.2 Inputs and outputs of the *single-* and *two-class* models, for the original (`Base`) and the upgraded (`Up`) systems. The two-class workload is $\mathbf{N} = \{255, 45\}$ jobs, $\boldsymbol{\beta} = \{0.85, 0.15\}$

		Single-class workload			Two-class workload			
		Aggregate			Adm		Doc	
		Base	Up wrong	Up correct	Base	Up	Base	Up
Inputs	D_{CPU}	0.180	0.036	0.039	0.2	0.04	0.1	0.02
	D_{Sto}	0.159	0.159	0.059	0.05	0.05	0.6	0.6
	Z_0	4.390	4.390	3.118	3	3	10	10
	R_0	49.649	43.38	14.683	54.322	12.393	30.801	147.228
Output	X_0	5.551	6.279	16.857	4.448	16.565	1.102	0.2862
Measures	U_{CPU}	1	0.2262	0.6683	0.8897	0.6626	0.1103	0.0057
	U_{Sto}	0.8842	1	1	0.2224	0.8283	0.6617	0.1717

In what follows we will refer to the values reported in Table 3.2 obtained with the two-class workload $\mathbf{N} = \{255, 45\}$ jobs, the corresponding population mix is $\boldsymbol{\beta} = \{0.85, 0.15\}$. The study has been carried out according to the following steps:

step (1)—First, we assume to know that the workload consists of $N_0 = 300$ jobs belonging to two classes *Adm* and *Doc* whose service demands are shown in Table 3.1. We consider the population mix $\boldsymbol{\beta} = \{0.85, 0.15\}$, i.e., the workload is $\mathbf{N} = \{255, 45\}$ jobs, 255 *Adm* and 45 *Doc*. Some output measures (`System Response time` R_0, `System Throughput` X_0, and the `Utilizations` of CPU and Storage) obtained from the execution of the two-class model are shown in columns `Adm-Base` and `Doc-Base` of Table 3.2.

step (2)—From the outputs of the two class model we compute the correspondent *single class* aggregate model (column `aggregate-Base`). The aggregate values of `Utilization` and `System Throughput` X_0 are obtained summing the correspondent per-class indexes: $U_{CPU} = U_{CPU,Adm} + U_{CPU,Doc} = 0.89 + 0.11 = 1$ $U_{Sto} = U_{Sto,Adm} + U_{Sto,Doc} = 0.22 + 0.66 = 0.88 X_0 = X_{0,Adm} + X_{0,Doc} = 4.448 + 1.102 = 5.551$. For the `System Response time` R_0, according to the Little law $N_i = X_i R_i$, the per-class values must be weighted by the relative throughput:

$$R_0^{Base} = \frac{N_0}{X_0} = R_{0,Adm} \frac{X_{0,Adm}}{X_0} + R_{0,Doc} \frac{X_{0,Doc}}{X_0} = 49.649 \, sec \qquad (3.3)$$

This is the correct computation of the `System Response time` with *multiclass workloads*. The same weights must be applied to the computation of the aggregate service demands $D_{CPU}^{Base} = 0.180 \, s$ and $D_{Sto}^{Base} = 0.159 \, s$. The weights of the relative throughputs have also the following intuitive interpretation. The number of jobs of the two classes *Adm* and *Doc* executed in the interval T are $C_{0,Adm}$ and $C_{0,Doc}$, respectively. This means that in the log file of the system executing the two classes workload there will be $C_{0,Adm}$ times the value of $R_{0,Adm}$ and $C_{0,Doc}$ times the value of $R_{0,Doc}$. Thus, to compute the mean of all the R_0s we need to weight the two values with the respective times they appear in the file. And, dividing by T both the terms of the ratios $C_{0,Adm}/C_0$ and $C_{0,Doc}/C_0$ we obtain $X_{0,Adm}/X_0$ and $X_{0,Doc}/X_0$, that are the weights considered in Eq. 3.3.

step (3)—Now consider a new analyst who does not know anything about the system workload and builds a single-class model considering all measures of the log file as belonging to a single type of jobs. If he made the right computations, he will get the same values shown in the column `Aggregate-Base` for both input parameters and output measures. We note that these values are the same as those already computed in the previous step. Indeed, he, without being aware, automatically applies for their computation the correct weights described in **step 2**. At this point, the analyst has the *wrong certainty* that the implemented model is *correct*!

step (4)—Now we will use the two workload models (the one with two classes and the one with a single class) for the performance projections. An increase of *Doc* customers is expected in the near future. We want to evaluate the effect on the response time $R_{0,Doc}$ of *Doc* jobs that will be generated by an increase of the CPU speed. We assume to consider a multicore processor that for the workload considered increases by a factor of five the CPU speed.

The primary effect of this upgrade is the decrease of the CPU service demands $D_{CPU,Adm}$ from 0.2 to 0.04 and of $D_{CPU,Doc}$ from 0.1 to 0.02 (see D_{CPU} in columns `Adm-Up` and `Doc-Up`, respectively). The execution of the two-class model compute the indexes reported in the lower part of these columns. Applying the computations described in **step 2)** it is possible to derive the correct values of the aggregated single-class model corresponding to the two-class model (see column `Aggregate-Up correct`).

Analyzing the `System Response times` of the two classes, we may see that while the $R_{0,Adm}$ decreases of 77%, the $R_{0,Doc}$ **increases of 377%** (from 30.8 to 147.2 s)! This is a **counterintuitive result: performance degrades with a CPU upgrade!** The motivation of this result hampered by intuition is related to the *switch* of the bottleneck from the CPU to the storage which take place after the upgrade. Indeed, since the throughput $X_{0,Adm}$ of class-*Adm* (that has the CPU as class-bottleneck) increases from 4.4 to 16.5 j/s, the competition for the *Storage* increases a lot (U_{Sto} reach saturation) and the class-*Doc* jobs, that are heavily storage bound (it is $D_{Sto,Doc} = 0.6\ s$, the Storage is the class-*Doc* bottleneck), experience a *strong degradation* of the response time.

Figure 3.5 shows the behavior of the `System Response times` of the two classes and the aggregated value as a function of all the population mixes, of the original system (a) and the upgraded version (b). The workload consists of 300 jobs, ranging from $\mathbf{N} = \{300, 0\}$ to $\mathbf{N} = \{0, 300\}$ jobs. The differences between the behavior of the correspondent curves are evident.

step (5)—We consider here the single-class model built in **step 3)** by the ignorant analyst, assuming that all the jobs as belonging to the same single class. The effect of the CPU upgrade is a decrease in service demand D_{CPU} from 0.180 to 0.036 s,

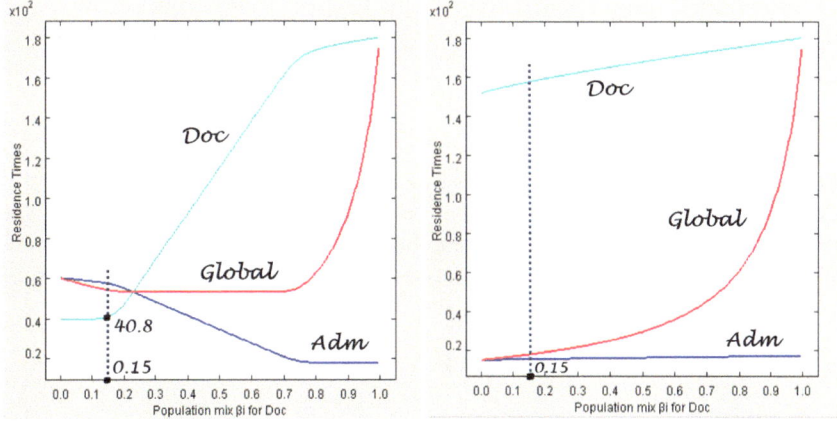

(a) SRT *vs* program mixes of Base system   (b) SRT *vs* program mixes of Up system

Fig. 3.5 System Response times *versus* all $\boldsymbol{\beta}$ population mixes of the *original* (Base) and the *upgraded* (Up) systems, with 300 jobs. Dashed lines represent $\boldsymbol{\beta} = \{0.85, 0.15\}$ used in Table 3.2

see columns (Aggregate-Base) and (Aggregate-Up wrong). The output measures (Table 3.2) obtained from the execution of this single class model with $N = 300$ *jobs* show an improvement of performance: R_0 *decreases* from 49.64 to 43.38 *s*, and X_0 *increases* from 5.5 to 6.2 *j/s*. As this qualitative behavior corresponds to the expectations, the analyst my have the wrong impression that the implemented model is correct. Indeed, comparing the values of the two columns Aggregate-Up wrong and Aggregate-Up correct it is possible to understand immediately the **large errors** affecting R_0 and X_0: -66% and +168%, respectively. Thus, we may conclude that

> the performance forecast based on single-class models of heterogeneous workloads are inaccurate when used to obtain projections for the average aggregate job. Furthermore, as Fig. 3.5 shows, it is evident that the average aggregate job cannot be used to derive reliable projections for each class of the two-class workload.

Let us remark that the per-class System Response times plotted by JMVA in Fig. 3.5 are the sum of the Residence times of a job of that class at *all* the resources of the system, *including* the Reference station. Thus, for example, we should add $Z_{0,Doc} = 10$ s (the Residence time of Doc jobs at the Reference station) to the System Response time of Doc jobs $R_{0,Doc} = 30.801$ s of Table 3.2 to obtain the value 40.801 plotted in Fig. 3.5a in correspondence to program mix $\boldsymbol{\beta} = \{0.85, 0.15\}$.

3.3   Performance Optimization of a Data Center

tags: closed, two-class, Queue, JMVA.

In this case study we illustrate a general approach applicable to several capacity planning problems with an example based on the performance optimization of a data center with a workload consisting of *heterogeneous* applications, [14, 38]. The problems of bottleneck identification and migration are addressed as a function of the fluctuations of the different types (*classes*) of requests being executed. The topic of *load balancing* is also investigated.

3.3.1   Problem Description

A data center partition consists of six servers utilized by business critical applications that access to sensitive data stored. The area is highly protected for both physical access and digital security. It consists of one `Web Server`, two `Application Servers`, and three `Storage Servers` (see Fig. 3.6). The access to the applications and data stored on these servers is permitted only to a limited number of employees with the appropriate authorization. Based on the different requirements, in terms of amount of resources used and Quality of Service (QoS) targets, *two* types of applications can be identified in the workload. The *two classes* of requests, called class-1 and class-2, generated by the two applications are focused on business logic processing the first, and on intensive data processing (search, update) the second.

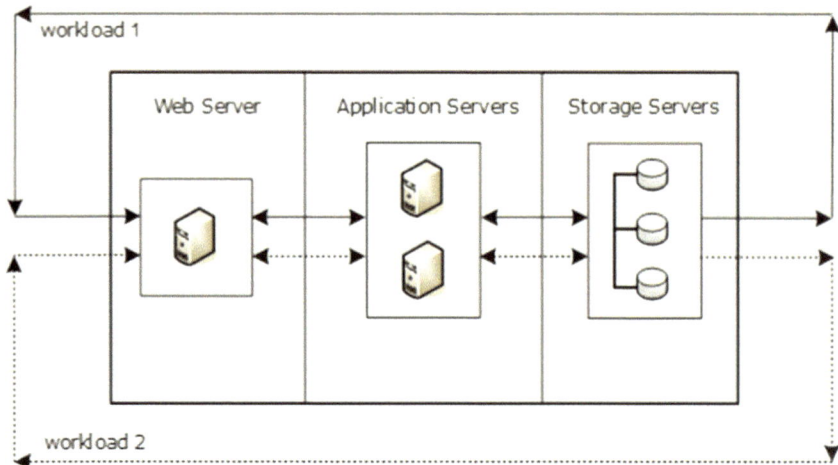

Fig. 3.6 Layout of the data center considered

The mean number of their requests that are in execution simultaneously is 100. According to forecasts, the number of employees authorized to access the servers is expected to double over the next nine months. The management is concerned that performance may degrade to an unacceptable level. Several initiatives are considered.

As a first action it is required to investigate the impact on performance of an increase of the global number N_0 of requests in execution from 100 to 200. More precisely, it is required to know if the QoS in terms of the mean per-class Response times defined for 100 employees are still satisfied. Since it is not possible to know how the fractions of the requests in execution of the two classes vary over time, to compute the upper bound of Response times it is necessary to consider all the possible combinations of the two classes of requests in execution, i.e., all the *population mix*, with $N_0 = 200$ req.

The capacity planning study should answer to several questions such as "Does the data center with the current configuration support the increase of the workload without saturating one or more resources?", "With $N_0 = 200$ *req* in execution will the QoS targets on the per-class Response times be satisfied?", "Which is the resource that is the bottleneck of the system?".

Other important questions to be answered concern the actions that should be taken to increase, if possible, the performance of the data center with the *current* configuration: "Which is the impact of the population mix on the potential increase of performance?", "The utilization of the resources are balanced?", "Which is the population mix that *maximizes* the System Throughput and *minimize* the mean System Response time (referred to as *optimal population mix*)?".

These questions have been grouped into the two *Objectives* of the study that are analyzed below.

3.3.2 Model Implementation

We need to evaluate the performance of the data center with an overall number of customers doubled (*200*) with respect to the current one (*100*). We implement a closed queueing model with six queue stations (see Fig. 3.6) and $N_0 = 200$ req in execution. The workload $\mathbf{N_0} = \{N_{0,1}, N_{0,2}\}$ consists of two classes of requests, where $N_{0,1}$ and $N_{0,2}$ are the number of requests in execution of *class*1 and *class*2, respectively.

To characterize the two type of applications in terms of processing requirements, a set of Service demands, one for each resource and each class, is used. The Service demand $D_{r,c}$ of a request of class-c at resource r is the total amount of time the request requires at that resource in order to be completely executed. The $D_{r,c}$ are computed *ignoring contention* by other requests and may be estimated measuring utilizations and throughputs and using the equation $U_{r,c} = X_{0,c} D_{r,c}$, where $X_{0,c}$ is the system throughput for class-c requests and $U_{r,c}$ is their utilization of resource r. To minimize the errors in the *parameter estimation*, it is *recommended*, when possible, to collect the measurements when the two types of appli-

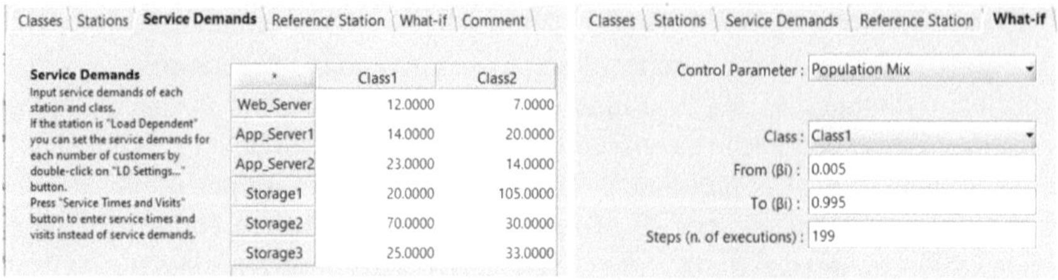

| Classes | Stations | **Service Demands** | Reference Station | What-if | Comment |

Service Demands		Class1	Class2
Input service demands of each station and class.			
If the station is "Load Dependent" you can set the service demands for	Web_Server	12.0000	7.0000
each number of customers by double-click on "LD Settings..."	App_Server1	14.0000	20.0000
button.	App_Server2	23.0000	14.0000
Press "Service Times and Visits" button to enter service times and	Storage1	20.0000	105.0000
visits instead of service demands.	Storage2	70.0000	30.0000
	Storage3	25.0000	33.0000

| Classes | Stations | Service Demands | Reference Station | **What-if** |

Control Parameter : Population Mix

Class : Class1

From (βi) : 0.005

To (βi) : 0.995

Steps (n. of executions) : 199

(a) Service Demands of the original workload   (b) All mixes with the two classes

Fig. 3.7 Service demands of the two classes of requests (**a**), and What-if parameters (**b**)

cations are executed in isolation. The Service demands of the two classes of requests, in *ms*, are shown in Fig. 3.7a. The amount of work requested from the Web Server is much less demanding than the one requested from the Application and Storage Servers. The computations required by the business logic place a medium load on the Application Servers while the high number of data manipulated, uploaded and downloaded, generate a high load on the Storage Servers.

The Service demands are exponentially distributed and the scheduling discipline of the servers is processor sharing PS. These assumptions allows us to solve the model analytically with the MVA algorithm [25, 31] using the JMVA. Indeed, according to the BCMP theorem (see Sect. 3.1), multiclass models in which the queue stations adopt the PS scheduling discipline can be solved analytically with efficient algorithms also if the service times (as may be considered the service demands of the single visit used our case) of the two classes at the same resource are *different*. Models in which these assumptions are not satisfied (e.g., if FCFS scheduling is required), must be solved with the simulation technique using JSIM.

The performance predictions obtained by the capacity planning study are based on several What-if analyses.

To evaluate the performance metrics corresponding to all the possible population mix we used the What-if feature of JMVA varying from 100% to 0% the requests in execution of one class and the opposite (from 0% to 100%) the fraction of the other class.

3.3.3   Results

The peculiarity of the *workload forecasting strategy* adopted in this study is that *only* the *global intensity*, i.e., the total number $N_0 = 200$ requests in execution is known, and it is *not* possible to predict how the fractions of the requests of the two classes (i.e., the *population mix*) vary over time. Typically, it may be quite bursty.

Thus, to achieve the capacity planning goals `What-if` analyses are required with `population mix` as control parameter, Fig. 3.7b. The fraction β_1 of class-1 requests in execution range from 0.5% to 99.5%, the one of class-2 is the complement to 100%.

Obj.1: Evaluate the behavior of the performance indexes of the data center with a global `Number of requests` in execution $N_0 = 200$ and for all the possible combinations of the two classes.

JMVA is used to estimate the performance of the data center for the required parameter range with the `Service demands` shown in Fig. 3.7a. In the `What-if` screenshot of Fig. 3.7b the control parameter is `population mix`, and 199 models are executed with *all* the possible mix of the requests of the two classes ranging from $\mathbf{N_0} = \{1, 199\}$ to $\mathbf{N_0} = \{199, 1\}$.

Let's start with the analysis of the behavior of `per-class` and `Global System Response` times with respect to all the population mix with $N_0 = 200$ req shown in Fig. 3.8a. The x-axis represents the fraction of class-1 requests β_1 with respect to the total number of requests in execution. In the two extremes $\beta_1 = 0$ and $\beta_1 = 1$ the workload consists of a *single* class, class-2 and class-1 only, respectively. In these cases, the resource that limit the performance of the system, i.e, the bottleneck, corresponds to the one with the max service demand D_{max}. When the bottleneck is saturated ($U_{bott} = 1$), the values of `System Response` times can be easily computed considering the D_{max} and applying the *Utilization* and *Little* laws, see Sect. 1.2. With **only** class-2 requests ($\beta_1 = 0$), the bottleneck is `Storage1` with $D_{Sto1} = D_{max} = 0.105$ s, from the *Utilization* law $U_{Sto1} = X_0 D_{max} = 1$ we derive

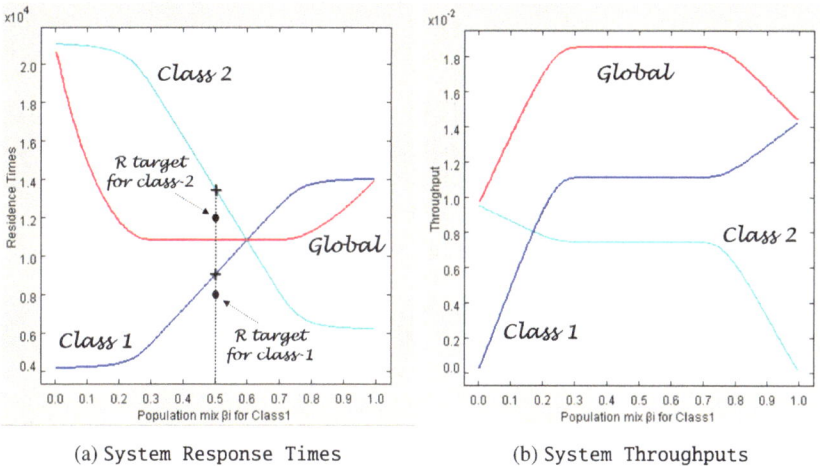

(a) `System Response Times`   (b) `System Throughputs`

Fig. 3.8 `System Response` times [ms] (**a**) and `System Throughputs` [*req/s*] (**b**) with $N_0 = 200$ req *versus* fraction of class-1 requests in execution, from 0.5% to 99.5%

$X_0 = 1/D_{max}$ and by *Little* law it will be $R_0^{max} = N_0 D_{max} = 200 \times 0.105 = \mathbf{21}$ s. In the other extreme $\beta_1 = 1$ with **only** class-1 requests, the bottleneck is Storage2 with $D_{Sto2} = D_{max} = 0.070$ s and it will be $R_0^{max} = N_0 D_{max} = 200 \times 0.070 = \mathbf{14}$ s.

As soon as the number of class-1 requests in execution increases (and thus class-2 requests decrease), the load of Storage2 starts to grow. As a consequence, the bottleneck tends to migrate from Storage1, i.e., the class-2 bottleneck, to Storage2, i.e., the class-1 bottleneck. The corresponding class-1 System Response time $R_{0,1}$ increases until 13.86 s with the workload $\mathbf{N_0} = \{199, 1\}$, very close to the asymptotic value 14 s above computed. It does not coincide with it because with 199 req of class-1 and 1 req of class-2 the Utilization $U_{Sto2,1}$ of Storage2 for *class-1 req.* is 0.995 and not 1. Similar motivations apply with the opposite workload $\mathbf{N_0} = \{1, 199\}$, i.e., $\beta_1 = 0.005$, where it is $U_{Sto1,2} = 0.995$ and the class-2 System Response time $R_{0,2}$ is 20.85 s while the asymptotic value is 21 s.

As can be seen in Fig. 3.8a, the Global System Response time R_0 is practically constant for a wide range of mixes, approximately between 30% and 70% of class-1. The important feature of these mixes is that executing a workload with one of them will result in the saturation of two resources *simultaneously*.

This interval of joint saturation, referred to as *common saturation sector*, is important in order to find the load of the system that optimize the performance, i.e., Throughput maximization and Response time minimization.

The identification of this interval can be done analytically under the assumption that the workload in execution is very large so that the bottleneck(s) is saturated [3]. With our workload of *200* req. the extremes of this interval are only approximate since the load does not saturate completely the bottlenecks.

The Response times of the two classes are identical when the two bottlenecks are *equiloaded* (for $\beta_1 = 0.6$ it is $R_0, 1 = R_0, 2 = 10.8$ s) and it can be shown that the corresponding *equiload* mix lies *inside* the common saturation sector [35]. R_0 is *minimum* in correspondence to this mix.

Figure 3.8b shows the System Throughput, Global X_0 and per-class $X_{0,1}$ $X_{0,2}$. It is evident that X_0 is *maximized* for all the mix of requests that belong to the common saturation sector while the per-class throughput are *constant* in the interval. It can be shown that the *equiutilization* point, i.e., the mix of the two classes that causes two bottlenecks to be equally utilized, lies into this interval and provides the *optimal load* that *maximizes* [35] the global System Throughput.

The behavior of Response times and Throughputs in Fig. 3.8 can be understood by analyzing the bottleneck *migration*. Indeed, it is known that the resource that limit the performance of the system under all possible workload mix is the bottleneck. So, when the bottleneck changes also the performance changes.

The Utilizations of the three Storage Servers of Fig. 3.9a show graphically this phenomenon. We do not consider Web and App servers because their Service demands, see Fig. 3.7a, are definitively lower than those of Storage servers, and therefore will never be saturated. As predicted, Storage1 and Storage2 saturate *together* for all the mixes of the common saturation sector (approximately between 30% and 70% fractions of class-1 *req.* of the total population of 200 req.) while the Utilization of Storage3 is definitively lower (its max Utilization reached in the common saturation sector is $U_{Sto3} = 0.52$), since its Service demands are smaller with respect to those of Storage1 and Storage2. When only a few class-1 requests are in execution ($\beta_1 < 0.3$), the bottleneck is Storage1. On the other side, when the fraction of class-1 req in execution is high ($\beta_1 > 0.8$) *only* Storage2 is saturated.

With regard to the *Quality of Service* targets on the per-class System Response times that were set for the original configuration of the data center with a workload of $N_0 = 100$ req evenly divided between the two classes, i.e., $\beta = \{0.5, 0.5\}$, we can see that with $N_0 = 200$ req *cannot be satisfied*. Indeed, the *target* values assigned to the mean System Response times of the two classes were $R_{0,1} = 8$ s and $R_{0,2} = 12$ s, respectively. With $N_0 = 100$ req these values **were met** ($R_{0,1} = 4.5$ s and $R_{0,2} = 6.7$ s), with $N_0 = 200$ req otherwise they **are not** ($R_{0,1} = 9$ s and $R_{0,2} = 13.5$ s), see Fig. 3.8a. We will see in the following *Obj.2* that a load balancing action allows the satisfaction of the targets.

Obj.2: Investigate on the actions that may improve the performance of the data center with the current configuration and a workload of $N_0 = 200$ req. Can the System Response time objectives be achieved after these actions?
Figure 3.9a emphasizes the problem: the unbalanced utilization of the Storage Servers.

To enhance the performance we need to take into consideration the most heavily loaded servers, namely Storage1 and Storage2, since reducing the Service demands at resources other than the *bottlenecks* produces only *marginal* improvements. The *total* load on the two Storage Servers 1 and 2 is fairly balanced, $D_{Sto1} = 125$ ms and $D_{Sto2} = 100$ ms, while the load of the third server is much

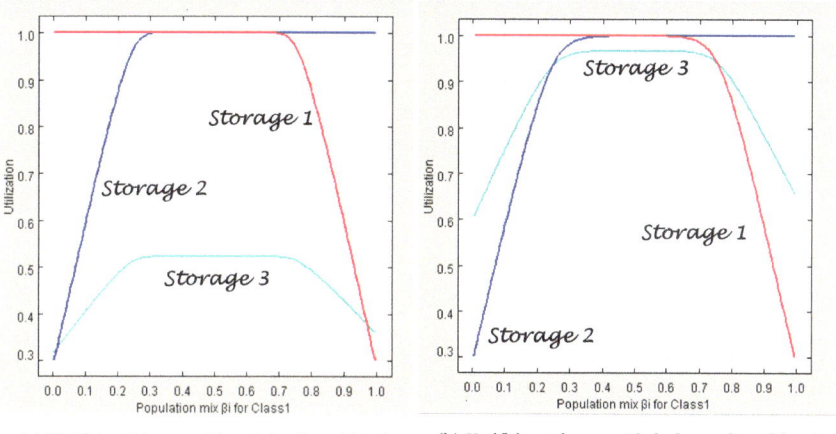

(a) Utilizations with *original* workload   (b) Utilizations with *balanced* workload

Fig. 3.9 Utilizations of the Storage servers in the *original* configuration (**a**) and in the *balanced* system (**b**) vs population mixes. The workload ranges from {0,200} to {200,0} req

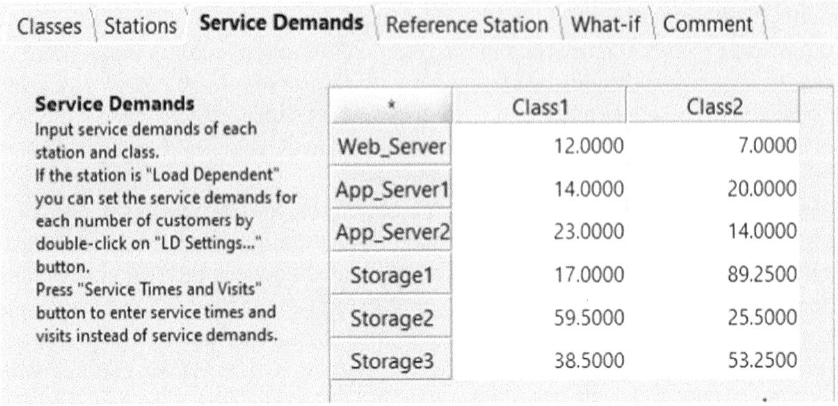

Classes \ Stations \ **Service Demands** \ Reference Station \ What-if \ Comment

Service Demands	*	Class1	Class2
Input service demands of each station and class.	Web_Server	12.0000	7.0000
If the station is "Load Dependent" you can set the service demands for	App_Server1	14.0000	20.0000
each number of customers by double-click on "LD Settings..."	App_Server2	23.0000	14.0000
button.	Storage1	17.0000	89.2500
Press "Service Times and Visits" button to enter service times and	Storage2	59.5000	25.5000
visits instead of service demands.	Storage3	38.5000	53.2500

Fig. 3.10 Service demands for the *balanced* configuration of the three Storage Servers

smaller, 58 ms, see Fig. 3.7a. We want to assess the effect of alleviating the bot-tlenecks, trying to balance the loads of all three Storage Servers. So, even according to usage statistics, some files have been migrated between the various storage, more precisely from Storage1 and Storage2 to Storage3, in order to make their total Service demands more similar to each other than in the original configuration. Figure 3.10 shows the new Service demands.

Let us remark that the Global Service demand to all the three Storage Servers must remain the same as the one of the original configuration, namely 283 ms, since all the servers have the same technical characteristics. Thus, shifting some data from one resource to another alter the visits V_r but not the Service time S_r of a request.

This configuration denotes a good balancing of the load of the three storages and no server is underutilized, Fig. 3.9b. By comparing their behavior with those of Fig. 3.9a obtained with the original system it is evident that the sum of the three U_r is *maximized*, thus enabling the maximization of the System Throughput.

The *maximum* System Throughput of the *original* system, obtained with fractions inside the common saturation sector, was $X_0^{max} = 0.0185$ req/ms, Fig. 3.8b, while the one obtained after the *balancing* action, Fig. 3.11b, is $X_0^{max} = 0.0218$ req/ms, with an *improvement* of about **17.8%**.

The corresponding *minimum* System Response time were $R_0^{min} = 10.8$ s for the *original* system, Fig. 3.8a, and $R_0^{min} = 9.18$ s for the *balanced*, Fig. 3.11a, with a *reduction* of about **15%**.

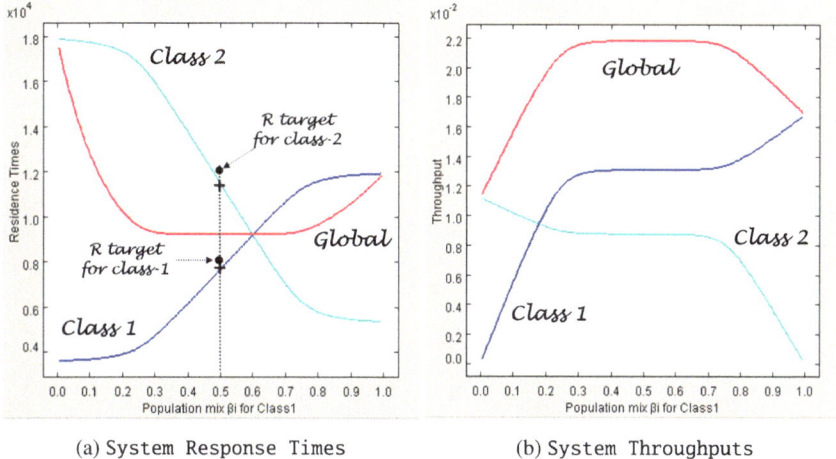

(a) System Response Times   (b) System Throughputs

Fig. 3.11 System Response times [*ms*] and Throughputs [*req/ms*] in the *balanced* configuration *versus* population mixes. The workload ranges from {0,200} to {200,0} req

With regard to the targets defined for the per-class System Response times, i.e., $R_{0,1} = 8$ s and $R_{0,2} = 12$ s, with the mix $\boldsymbol{\beta} = \{0.5, 0.5\}$, we can see from Fig. 3.11b that the *balanced* configuration is **able to satisfy** them, i.e., $R_{0,1} = 7.65$ s and $R_{0,2} = 11.47$ s. Instead, as described previously, the original configuration of data center with 200 req in execution was **unable** to reach them.

Chapter 4
Impact of Variability of Interarrival and Service Times

4.1 Importance of Distributions: A Motivating Example

In this section we highlight the *significant errors* in the computation of performance indexes that are introduced when *only* the mean values are considered instead of the distributions of some input parameters. Consider a server that requires a *constant* Service time S of 1 s to execute a request. Assume that the requests arrive with rate $\lambda = 60$ req/min in groups (*bursts*) and that the requests of a burst arrive at the same instant of time. The time between consecutive bursts is *constant*. We will analyze the impact on Queue time and Response time of different burst lengths, ranging from 1 to 60, considering always the *same* arrival rate.

In the case shown in Fig. 4.1a, a request arrives at the server exactly every *second*. Since the time S required for its execution is always equal to 1 s, the queue will *never* take place (the Queue time is equal to zero) and thus the mean Response time (Queue time plus Service time) is exactly one *second* for all *requests*. In the other graphs it is assumed that the requests arrive at the server with burst of increasing dimensions.

In Fig. 4.1b a burst of size 2 arrives exactly every two *seconds*. The first request never waits in queue, while the latter waits for one *second*, that is, the execution time of the first. So the mean Queue time is 0.5 s. In the graph of Fig. 4.1c a burst of size 3 arrives exactly every three *seconds*. The first request never waits, the second waits a *second* and the third waits two *seconds*. So the mean Queue time is 1 s and the mean Response time is 2 s.

Finally, in Fig. 4.1d 60 *requests* arrive together in a single burst every sixty *seconds* (the rate is always 1 req/s). In this case the mean Queue time is 29.5 s. Let us remind that the sum of n positive consecutive integers starting from 1 is $n(n + 1)/2$. In our case we have 60 requests, but *only* $n = 59$ of them wait from 1 to 59 s, respectively. Thus, the *mean* waiting time (Queue time) of the 60 requests is 29.5 s and the mean Response time is 30.5 s! The conclusion is

© The Author(s) 2024
G. Serazzi, *Performance Engineering*,
https://doi.org/10.1007/978-3-031-36763-2_4

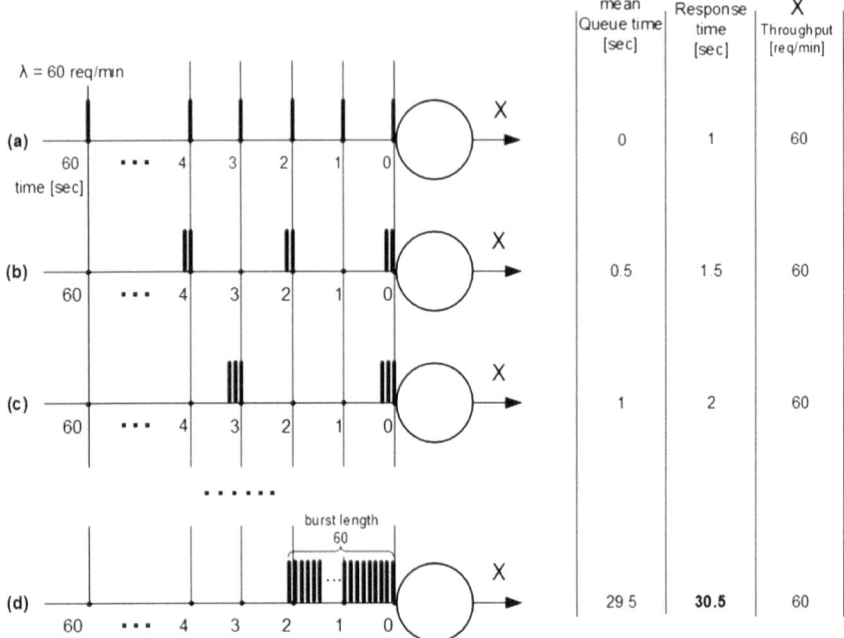

Fig. 4.1 Impact of different burst lengths on mean Response time

Even considering the **same** *arrival rate* $\lambda = 1$ req/s *and the* **same** Service times $S=1$ s, *depending on the arrival pattern of requests we could have a* **very high variability** *of mean* Response times: *from 1 to 30.5 s in the example considered (and this is not the worst case!).*

4.2   Variability of Interarrival Times

tags: open, single class, Queue, Exp/Hypo-exp/Hyper-exp, JSIMg.

The objective of this case study is to emphasize the impact of the variance of Interarrival times on the performance of a system.

4.2.1   *Problem Description*

Consider a model of a web server that needs to execute an e-commerce application to sell equipment produced by a new company. While the mean and variance of Service time required to process a purchase order can be estimated with sufficient accuracy, the pattern of incoming requests is unknown as customers are located all over the world.

We use a simple model of the web server consisting of a single queue station. To account for the unknown patterns of the incoming requests we consider *five* distributions of interarrival times with the *same mean* and increasing variability. To describe the variance of interarrival times we use the *coefficient of variation* c of each distribution (given by the *standard deviation/mean* ratio). For a given value of arrival rate, the values of c are directly proportional to the variance of the five distributions since their means are the same.

The Service times are assumed *exponentially* distributed with the same mean $S = 1$ s for *all* the models.

To analyze a wide range of traffic intensities we consider several arrival rates, ranging from light-load (10% of server utilization) to heavy-load (90% of server utilization) conditions. For *each* arrival rate we execute *five* models corresponding to the *five* interarrival time distributions. As a reference metric we consider the mean Response times of the models executed. The models are solved with JSIMg.

4.2.2   Model Implementation

We use a open model consisting of three stations: Source1, Queue1, and Sink1, Fig. 4.2a. The Service times of Queue1, with mean $S = 1$ s, used in all the models have the same *exponential* distribution. The five distributions considered of Interarrival times, in sequence of increasing variance are: *Constant* cv = 0, *Hypo-exponential* cv = 0.5 (Hypo-exp), *Exponential* cv = 1 (Exp), *Hyper-exponential* cv = 5 (Hyper-exp), *Hyper-exponential* cv = 10 (Hyper-exp). Figure 4.2b shows the window for setting the mean = 10 (corresponding to $\lambda = 0.1$ req/s) and the coefficient of variation cv = 10 of the *Hyper-exp* distribution.

The differences between the distributions are emphasized in Fig. 4.3a (obtained with $\lambda = 0.9$ req/s), that shows the graphs relating to three of them: *Hypo-exp* cv = 0.5, *Exp* cv = 1, and *Hyper-exp* cv = 0.5. As can be seen, the percentages of Interarrival times (i.e., the *percentiles*) that are *less than* the mean value 1.111 s are very different: 56.8% for the *Hypo-exp*, 63.6% for the *Exp* (the exact analytical result is 0.6321), and 85% for the *Hyper-exp* with cv = 5 (and 91% for the *Hyper-exp* cv = 10, not shown in the figure). To obtain the percentiles of a metric with JSIMg see Sect. 2.2 and Figs. 2.10, 2.11.

The increase of variability also heavily influences the *maximum* values of the various distributions: 5.4 s for the *Hypo-exp*, 16.69 s for the *Exp*, 261.22 s for the *Hyper-exp* cv = 5, and 864.46 s for the *Hyper-exp* cv = 10. The number of samples needed to reach the equilibrium of the metric Throughput of Source1, that provides the Interarrival times with 99% Confidence Interval and 0.03 Max Rel. Err., ranges from 40960 of the *Hypo-exp* to 1063920 of the *Hyper-exp* cv = 10.

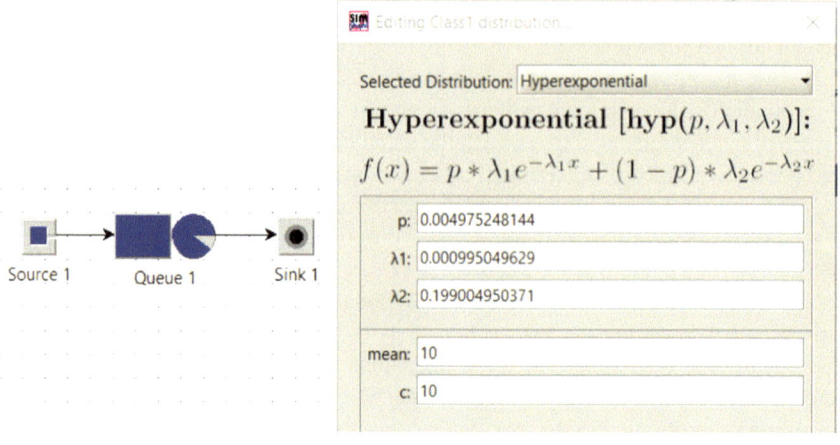

(a) Layout of the model   (b) Parameters of the Hyper-exp distribution

Fig. 4.2 The model considered (**a**), Settings of the mean = 10 and coeff. of variation cv = 10 of the *Hyper-exponential* distribution of Interarrival times for Arr.rate 0.1 req/s (**b**)

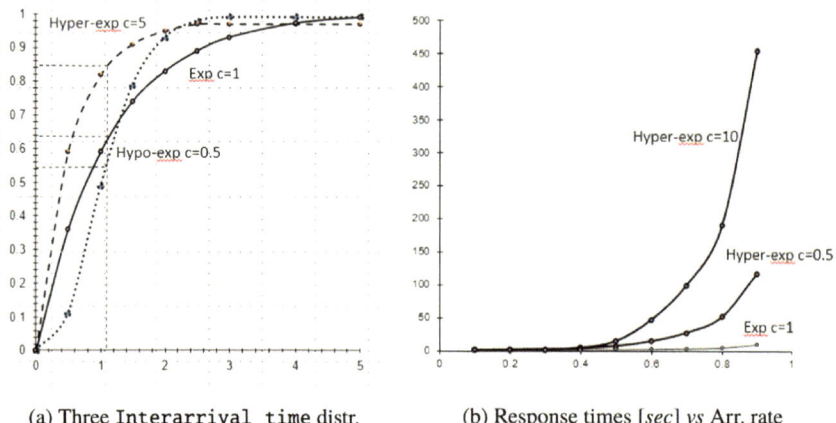

(a) Three Interarrival time distr.   (b) Response times [*sec*] *vs* Arr. rate

Fig. 4.3 Interarrival time distributions with *increasing variability* (cv = 0.5, 1, 5) obtained with λ = 0.9 req/s and the same mean 1.111 s (**a**); the corresponding Response times of Queue1 for λ = 0.1 ÷ 0.9 req/s (**b**)

Table 4.1 Response times [s] with five Interarrival time distributions with increasing variance vs Arrival rates. Service times S = 1 s are *exponentially* distributed

Arrival rate	Response times				
	Interarrival time distributions				
	Const cv = 0	Hypo-exp cv = 0.5	Exp cv = 1	Hyper-exp cv = 5	Hyper-exp cv = 10
$\lambda = 0.1$ [req/s]	1.00	1.01	1.11	1.22	1.24
$\lambda = 0.3$ [req/s]	1.05	1.12	1.43	2.20	2.40
$\lambda = 0.6$ [req/s]	1.47	1.70	2.54	14.49	46.98
$\lambda = 0.9$ [req/s]	5.13	6.43	9.92	116.88	455.06

4.2.3 Results

To simulate the different traffic intensities we use, for each distribution, a What-if analysis, with *Arrival rate* as control parameter, that execute nine models with λ ranging from 0.1 (light load) to 0.9 (heavy load) req/s with increments of 0.1. Figure 4.3b shows how the Response time R varies with different arrival patterns and rates. To make it easier to understand the figure, only R obtained with three distributions are plotted: *Exp* cv = 1, and *Hyper-exp* with cv = 5 and cv = 10. As can be seen, the values of R grow very fast not only when the Arrival rate is approaching the saturation value $\lambda^{sat} = 1$ req/s (and expected) but also with the increase of the variability of the Interarrival times (and this is not so expected).

Table 4.1 shows the Response Times for the five distributions with Arrival rates $\lambda = 0.1, 0.3, 0.6, 0.9$ req/s.

Even if we do not consider the two extreme distributions (i.e., the Constant cv = 0 and the Hyper-exp cv = 10), the differences between the Response times corresponding to the same λ become greater as the utilization of the server increases. The values of the last row of the table, corresponding to the utilization of 90%, show a difference of *more than 18 times* between 6.43 s with Hypo-exp cv = 0.5 and 116.88 s with Hyper-exp cv = 5!

Since for a given arrival rate λ the server *utilization U* is the *same* for *all* distributions (it is $U = \lambda S$), we may conclude that:

> *measuring server* Utilization *is* **useless** *to predict* Response *times if it is not complemented with the knowledge of other metrics, such as the distributions of* Interarrival *and* Service *times.*

4.3 Variability of Service Times

tags: open, single class, Queue, Exp/Hypo-Exp/Hyper-Exp, JSIMg.

This case study has been purposely designed to highlight the impact of the variance of `Service times` on the performance of a system. The `Service times` follow five different distributions, while the `Interarrival times` are generated according to the same *Exponential* distribution. It can be considered the dual of the example discussed in the preceding section in which the opposite situation was evaluated.

4.3.1 Problem Description

The scenario of this example is quite common in many practical problems in which the execution times of the applications are often *highly variable* based on input data and required functions (see, e.g., [21]).

We consider an application for the computation of the path between two geographical locations. The algorithms that compute the driving route from a source to a destination are computationally heavy and the `Service demands` are *highly variable* as a function of the locations considered. For these reasons the management decided to deploy the application on a dedicated server and to evaluate the impact on `Response time` of the different locations.

The `Interarrival times` of the route requests are assumed *Exponentially* distributed and different `Arrival rates`, that cover the range from light to heavy traffic, are considered. To account for the different fluctuations in execution times, *five* distributions with *increasing* variances, from zero to very high values, and the same mean were considered. For each `Arrival rate` we evaluate the `Response time` for the five `Service times` distributions. The models are solved with JSIMg.

4.3.2 Model Implementation

The layout of the open model used is shown in Fig. 4.4a. It consists of three stations: `Source1`, `Queue1`, and `Sink1`. The five distributions of the `Service times` considered, in sequence of increasing variance, are: *Constant* cv $= 0$ (`Const`), *Hypo-exponential* cv $= 0.5$ (`Hypo-exp`), *Exponential* cv $= 1$ (`Exp`), *Hyper-exponential* cv $= 5$ (`Hyper-exp`), *Hyper-exponential* cv $= 10$ (`Hyper-exp`). The use the *coefficient of variation* cv of each distribution (given by the *standard deviation/mean* ratio) to describe the variance of `Service times` is convenient in this case as,

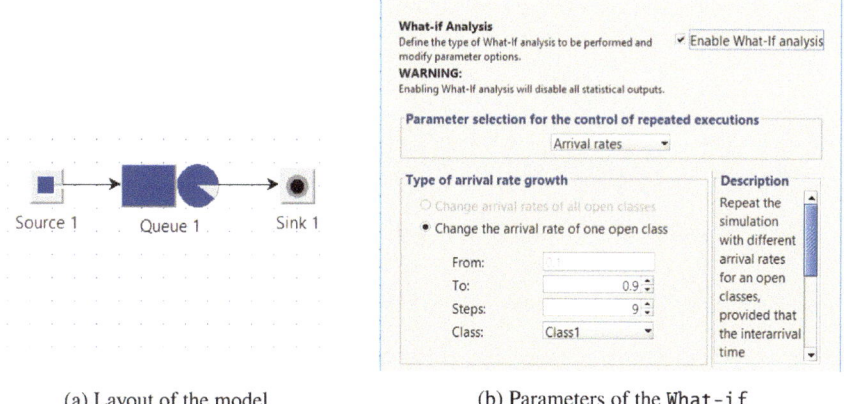

(a) Layout of the model   (b) Parameters of the What-if

Fig. 4.4   Model considered (**a**); What-if with Arrival rates λ = 0.1 ÷ 0.9 req/s (**b**)

for a given Arrival rate, its values are directly proportional to the variance of the five distributions being their means the same (S = 1 s).

The same *Exponential* distribution of the Interarrival times generated by Source1 is used in all the models. A What-if analysis is used to execute, for each distribution of Service times, 9 models with Arrival rates ranging from 0.1 to 0.9 req/s with increments of 0.1 (see Fig. 4.4b). Globally, five What-if analyses are required corresponding to the five distributions of Service times considered (in total 45 models are executed).

4.3.3   Results

The objective of the two graphs of Fig. 4.5 is to provide a visual evidence of the negative effects of service time *fluctuations* on Response times. In Fig. 4.5a the Service times of a period of three hours (simulated time) with a Hyper-exp cv = 5 distribution are shown. Remember that the mean is S = 1 s for all distributions! The Response times, with λ = 0.9 req/s, for the same period are shown in Fig. 4.5. The data for the plots of Fig. 4.5 are obtained from the CSV files generated by JSIMg.

The correlation between the bursts of high values of S and the peaks of Response times is evident and consistent with intuition. The *bursts* create a congestion of the server and small increases in arriving requests in this condition determine enormous increases in queue length, and in Response times together with it. For example, consider the initial period of half-hour, or the period of about 800 s centered at the end of two hours (7200 s), or the period starting at about 9000 s. It must be pointed out that the fluctuations of Response times are emphasized in our case due to the high Utilization of the server $U = \lambda S = 0.9$.

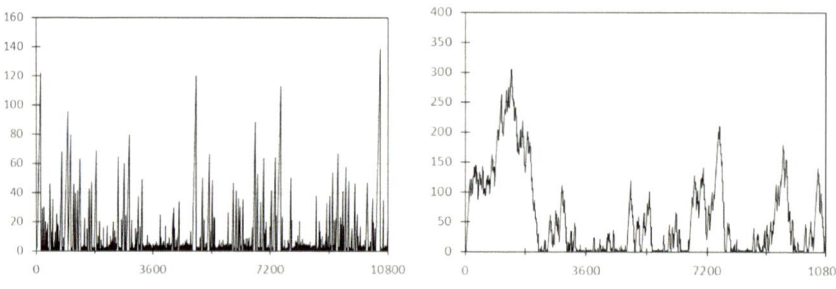

(a) Fluctuations of S *Hyper-exp* cv=5   (b) The corresponding `Response times` [*sec*]

Fig. 4.5 `Service times` generated with *Hyper-exp* distribution (S = 1 s and cv = 5) for a period of three hours (**a**); corresponding `Response times` with λ = 0.9 req/s (**b**)

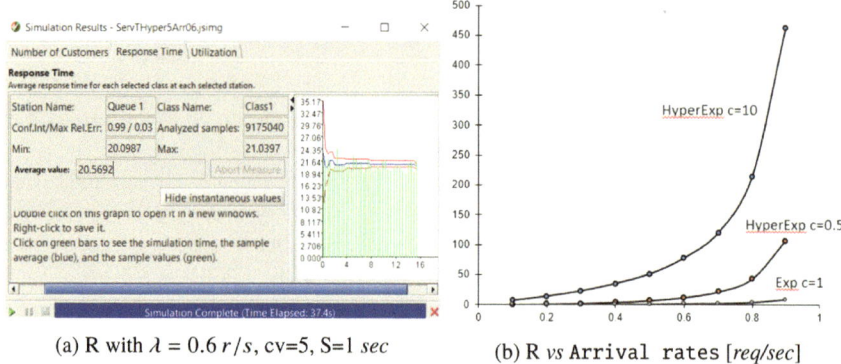

(a) R with λ = 0.6 r/s, cv=5, S=1 *sec*   (b) R *vs* `Arrival rates` [*req/sec*]

Fig. 4.6 `Response Time` with *Hyper-exp* cv = 5 distrib. of S (**a**); R with three different `Service times` distributions and same mean 1 s, `Interarrival times` are *Exponentially* distributed

Figure 4.6a shows an example of the results provided by one of the 45 models executed: the behavior of the `Response times` obtained from a simulation run with λ = 0.6 req/s and *Hyper-exp* distribution of `Service times` with cv = 5. The mean value R = 20.56 s with the precision required (99% of conf. interval, 0.03 max error) is obtained with 9175040 samples.

The `Response times` obtained with three different distributions of `Service times` are shown in Fig. 4.6b. The `arrival rate` range from 0.1 to 0.9 req/s with step of 0.1. The variance of the three distributions increases from the *Exponential* (cv = 1) to the *Hyper-exp* (cv = 10).

The `Response times` obtained by JSIMg simulating five distributions of `Service times` and λ = 0.1, 0.3, 0.6, 0.9 req/s are given in Table 4.2. As can be seen, for the same `Arrival rate` there are huge differences between the values obtained with the five distributions. These differences increase as server `Utilization` increases. Even avoiding to consider the `Constant` cv=0

Table 4.2 Response times with five Service times distributions with increasing variance and same mean S = 1 s vs Arrival rates. Interarrival times are *Exponentially* distributed

Arrival rate	Response time				
	Service time distributions				
Exp cv = 1	Const cv = 0	Hypo-exp cv = 0.5	Exp cv = 1	Hyper-exp cv = 5	Hyper-exp cv = 10
$\lambda = 0.1$ [r/s]	1.05	1.06	1.11	2.42	6.67
$\lambda = 0.3$ [r/s]	1.21	1.26	1.43	6.66	22.62
$\lambda = 0.6$ [r/s]	1.76	1.95	2.54	20.56	77.15
$\lambda = 0.9$ [r/s]	5.53	6.51	9.92	119.17	453.36
$\lambda = 0.9$ M/G/1	**5.5**	**6.625**	**10**	**118**	**455.5**

distribution, which provides a lower bound for all distributions, we can have enormous differences (up to **70 times** with $\lambda = 0.9$ req/s) between the Response times obtained with *Hypo-exp* cv = 0.5 (6.51 s) and those with *Hyper-exp* cv = 10 (453.36 s)! Let us remark that these differences occur even if the Utilization of the server is the same for all distributions.

Thus, we can conclude that:

> *to provide* **accurate** *performance forecast of a server it is essential to know the distributions of* Interarrival *and* Service *times, and not just their mean values and server* Utilization.

The model considered in this section could be solved analytically obtaining exact results. In fact it corresponds to a M/G/1 queue station (see the tutorial [32] and, e.g., [36]) having *Exponential* Interarrival times, i.e., the arrival process is Poisson (Markovian, M), Service times with *general* distribution (G) with given mean and variance, and a single server. The Response time of this station is given by:

$$R_{Queue1} = \text{waiting time in queue } W + \text{Service time } S = \frac{U S(1 + cv^2)}{2(1 - U)} + S \quad (4.1)$$

where U is the server Utilization ($U = \lambda S$), and cv is the *coefficient of variation* of the *general* distribution of Service times with mean S. Note that both the mean and the variance of Service times *must be known* to compute the coefficient of variation. In the last row of Table 4.2 are reported the exact Response times computed with Eq. 4.1. As can be seen, the values obtained with JSIMg are very close to the exact ones, and are all within the 99% confidence intervals.

Let us remark that when the Service times are *Constant* it is cv = 0 and the model is identified as M/D/1 (D stands for *Deterministic* Service times). Its waiting time W (computed with Eq. 4.1) is half of that obtained with an *Exponential* distribution (in a M/M/1 model with cv = 1). For example, as shown in the last row of Table 4.2 with *Constant* distribution it is W = 4.5 s while with Exponential it is W = 9 s (with λ = 0.9 req/s and S = 1 s). The waiting time W of an M/D/1 station is the *lower bound* for any M/G/1 station with the same S and Arr. rate.

Chapter 5
Parallel Computing

5.1 Synchronization of All Parallel Tasks

tags: open, single class, Source/Fork/Queue/Join/Sink, JSIMg.

5.1.1 Problem Description

The focus of this problem is on the use of `Fork` and `Join` stations for parallel computing and synchronization. The model is *open* and the workload consists of a *single class* of jobs.

We will consider the problem of modeling the execution of a job that at some point (i.e., at the `Fork` station) splits into several parts, referred to as *tasks*, that will be executed in parallel. The tasks may be instances of the same code processing different data set, or part of the code performing different computations. Each task can follow a different path through the resources between the `Fork` and the `Join`. When all tasks complete their executions, they are merged in the `Join` station and then, according to the synchronization policy, the job that generated them can continue its execution. This type of behavior is typical of many current applications, such as *Map/Reduce*, that alternate phases in which various instances of the code are generated and executed in parallel with phases that require their synchronization.

5.1.2 Model Implementation

We use a `Fork` station that when a job arrives generates four equal jobs, referred to as *tasks*, that will be executed in parallel, and a `Join` station to synchronize their executions. When all the executions are completed, the `Join` releases, i.e., *fire*, the job. The layout of the model is shown in Fig. 5.1. A `Source` station generates the flow

© The Author(s) 2024
G. Serazzi, *Performance Engineering*,
https://doi.org/10.1007/978-3-031-36763-2_5

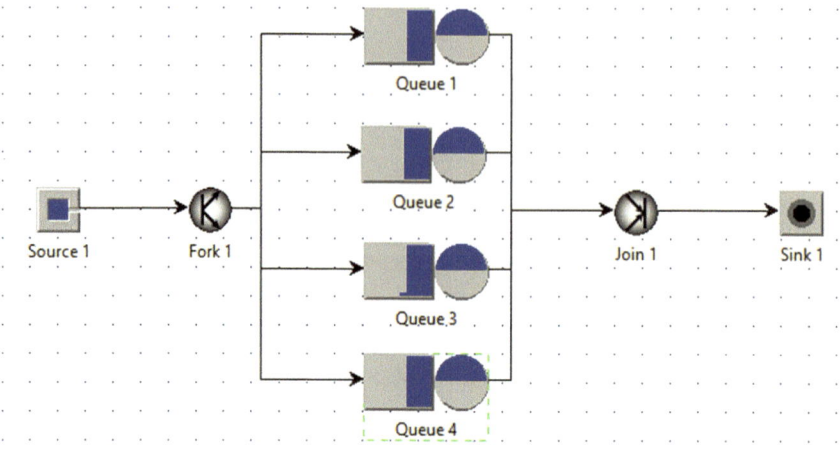

Fig. 5.1 Fork1 generates for each job four tasks executed in parallel and synchronized on Join1

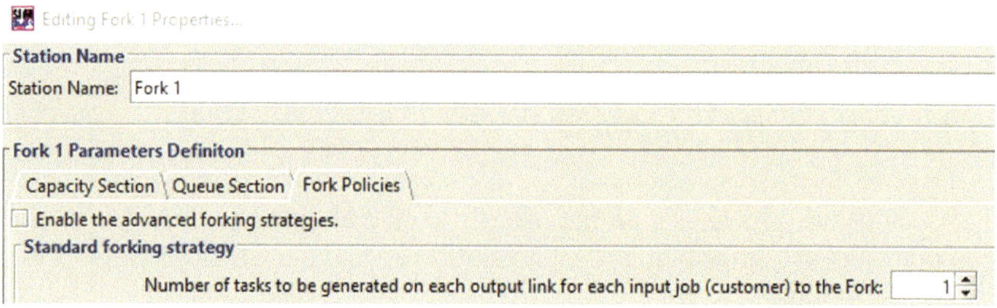

Fig. 5.2 Fork1 parameterization: one task is generated and sent on each output link

of jobs with exponentially distributed Interarrival times. The Service times S_i of the four queueing stations Queue1÷4 are *exponentially* distributed.

The Arrival rate of the jobs is $\lambda = 1$ j/s and the mean Service times of the four queue stations are $S_1 = S_2 = S_3 = S_4 = 0.5$ s.

In the Editing Fork Properties window (Fig. 5.2) we do not flag the check box for enabling the Advanced Forking Strategies so the Standard Strategy is applied.

For each arriving job, this strategy send *n tasks* on *each* link in output of the Fork. For *n* we left the default value $n = 1$. We want that all the four tasks generated by a job, one per output link, will be completely executed before the job exit the Join station. To *synchronize* all the tasks of a job at the Join station the Standard Join Strategy (see Fig. 5.3) is selected.

Initially we execute a model with Arrival rate $\lambda = 1$ j/s. The performance indexes to be collected together with the requested precision (in terms of confidence level and max relative error) of their values are shown in Fig. 5.4.

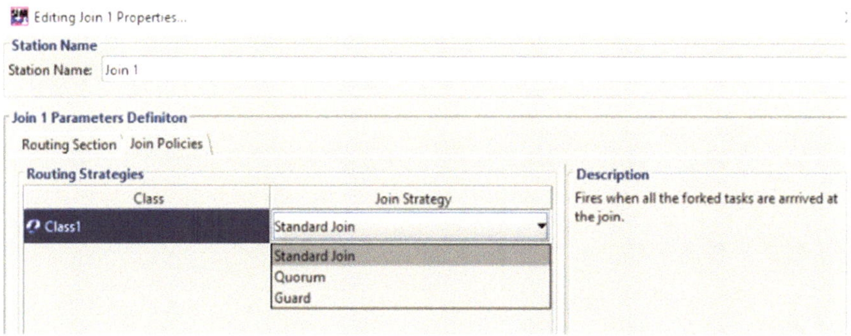

Fig. 5.3 `Join1` parameterization: executions of *all* tasks are synchronized before the job is released

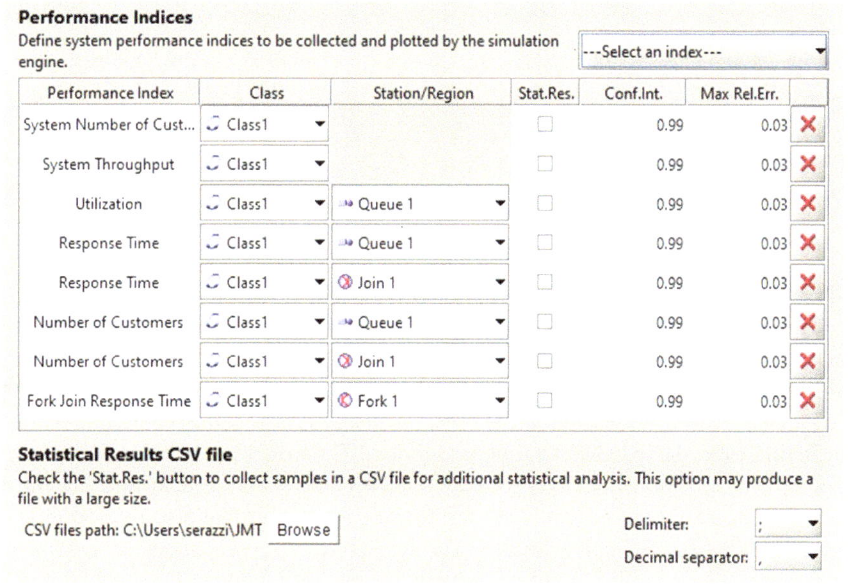

Fig. 5.4 Performance indexes collected during the simulation, and their precision parameters

After this single simulation run, we investigate the behavior of performance indexes for different values of arrival rate λ. To this end we use the `What-if` analysis feature (Fig. 5.5). We check the box `Enable what-if analysis`, and we select `Arrival rate` as *control parameter*. Five executions are required with arrival rates $\lambda = 1, 1.2, 1.4, 1.6, 1.8$ j/s.

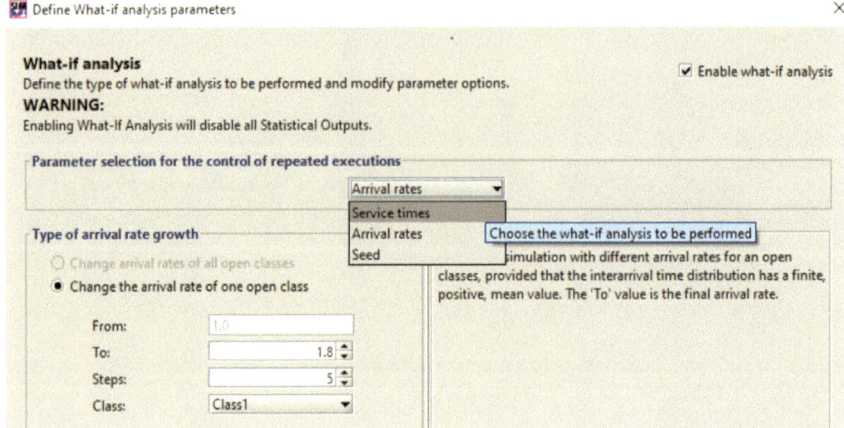

Fig. 5.5 `What-if` analysis: 5 models with `Arrival rates` from 1 to 1.8 j/s are executed

5.1.3 Results

In this section we show some of the results obtained from the simulations and we compare their values with the corresponding exact values computed analytically, when these are available.

The simulation with $\lambda = 1$ j/s provided the values of all the measured performance indexes with the precision required in Fig. 5.4.

In Fig. 5.6, the mean `Response times` of `Queue1` $R_{Q1} = 1.01$ s and of `Join1` $R_{J1} = 0.938$ s stations are shown. The `Response time` of `Queue1` is the mean time of a visit to `Queue1` (queue+service). The `Response time` of `Join1` is the *synchronization time* of the four tasks since it represents the mean time that three tasks, whose executions are already terminated, must wait that also the fourth end before the fire of the job can take place. The `Fork/Join Response time` (mean time within a `Fork/Join` section) provided by the simulation is $R_{FJ} = 1.92$ s and, in the model considered, is obtained by adding the *mean* of the four `Response times` of the queue stations and the `Synchronization time` of `Join1`.

The validation of the results of the individual queue stations, considered in isolation from the rest of the model, can be done by comparison with the corresponding exact values computed analytically. Indeed, each queue can be modeled as a M/M/1 station since both its `Interarrival times` and `Service times` are exponentially distributed. Thus, its `Utilization` is $U_i = \lambda S_i = 0.5$, its `Response Time` (mean time for one visit, queue plus `Service times`) $R_i = S_i/(1 - U_i) = 1$ s, its mean `Number of customers` (*tasks*) in the station $N_i = U_i/(1 - U_i) = 1\ task$.

The results obtained from the simulation are very close to these ones computed analytically: the `Response time` of `Queue1` is $R_{Q1} = 1.01$ s (Fig. 5.6), the `Utilization` is $U_{Q1} = 0.505$, and the *mean* `Number of tasks` is

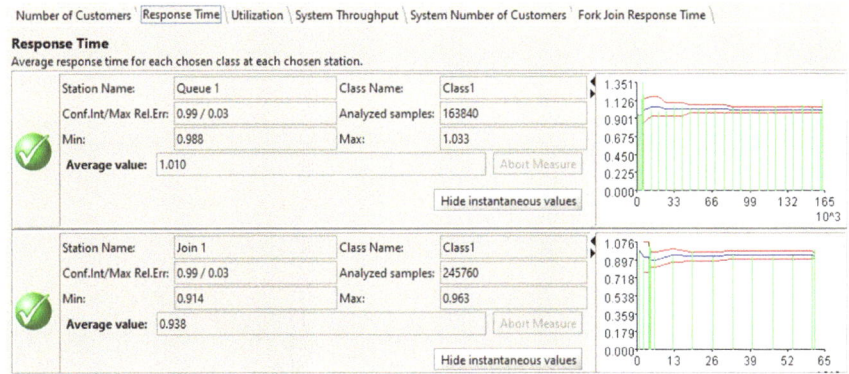

Fig. 5.6 Response times of Queue1 and Join1 stations (mean time for a visit to Queue1 and mean synchronization time at the Join1, respectively) for the model of Fig. 5.1 with $\lambda = 1$ j/s

$N_{Q1} = 1.01$ *tasks*. Similar values have been obtained for the other three stations Queue2, Queue3, and Queue4.

To study the behavior of the Fork/Join Response time, that includes the Synchronization time of the tasks at Join1, we use a What-if analysis (Fig. 5.5) requiring the simulation of *five* models with Arrival rates $\lambda = 1 \div 1.8$ j/s. The results are plotted in Fig. 5.7.

Unfortunately, the exact formula to compute the Fork/Join Response time is known only for particular models. In more general cases various approximate solutions are available.

The exact Fork/Join Response time can be computed only when there are *two* parallel paths in output of the Fork and the two servers are M/M/1 queue stations with the same service rate.

In this model, the *exact* mean Fork/Join Response time, see [17], is given by:

$$R_{FJ} = \frac{12 - U}{8} \frac{S}{1 - U} \tag{5.1}$$

and the exact mean Synchronization time at the Join (referred to as R_J) is

$$R_J = \frac{4 - U}{8} \frac{S}{1 - U} \tag{5.2}$$

The results obtained with simulation are validated considering the model of Fig. 5.8 whose exact Fork/Join Response time and Synchronization time are given by Eqs. 5.1 and 5.2, respectively. The parameters used are $\lambda = 1 \div 1.8$ j/s and $S_{Q1} = S_{Q2} = 0.5$ s and all the distributions are *exponential*.

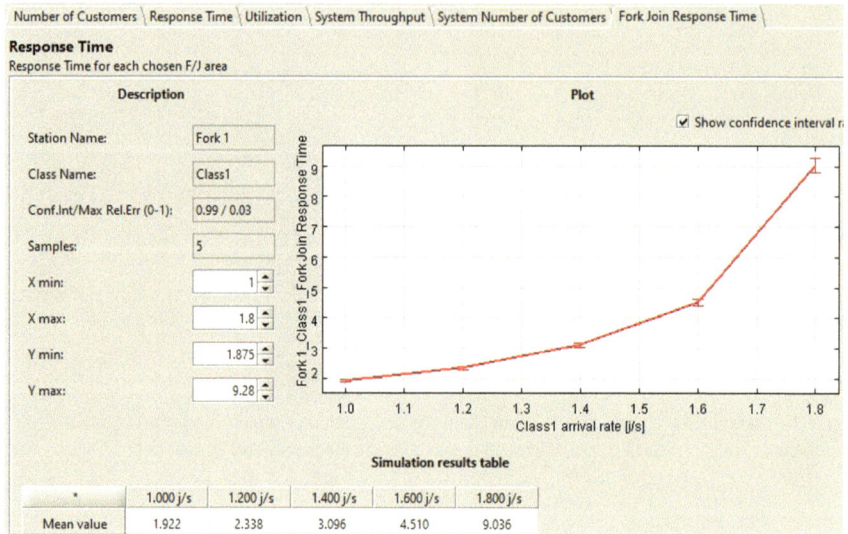

Fig. 5.7 Fork/Join Response time of the model of Fig. 5.1 computed with a What-if with the Arrival rate λ ranging from 1 to 1.8 j/s in five steps (see Fig. 5.5)

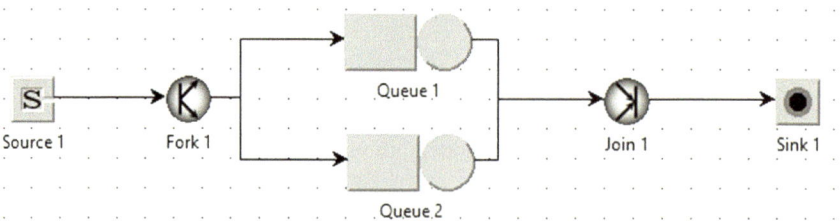

Fig. 5.8 Fork/Join with two equal Queues M/M/1 that can be solved analytically, Eqs. 5.1, 5.2

Table 5.1 shows the results obtained with JSIMg and the corresponding exact values. As can be seen, the exact values are within the 99% confidence intervals as required, (see Fig. 5.4).

For Fork/Join structures with a number of parallel paths *greater* than *two*, *heterogeneous* queue stations, and *general* distributions there are *no exact* formula to compute the performance indexes. However, several approximations, some enough precise but complex to compute, are available in literature (see, e.g., [28]).

An *estimation*, rather *coarse* but simple to compute, can be obtained considering the model typically adopted to study the reliability of *parallel infrastructures*. A system consisting of *n parallel components fails* when *all* the *n components fail*. Consider the instants in which the tasks complete their executions as events corresponding to the *failures* of components of the reliability model. We can see that the two models (the Fork/Join and the reliability) are similar since both seek the

Table 5.1 Fork/Join Response times and Synchronization times of the two equal parallel queues (Fig. 5.8) obtained with JSIMg and their *exact* values computed with Eqs. 5.1 and 5.2

Arrival rate [j/s]	Fork/join Response time		Synchronization time	
	JSIMg	Exact	JSIMg	Exact
$\lambda = 1.0$	1.413	1.437	0.431	0.437
$\lambda = 1.2$	1.775	1.781	0.526	0.531
$\lambda = 1.4$	2.378	2.354	0.692	0.687
$\lambda = 1.6$	3.478	3.5	0.981	1.0
$\lambda = 1.8$	6.971	6.937	1.956	1.937

mean time required for the *end* of *all* the n tasks or the *failures* of *all* the n components. In the reliability model several assumptions are typically made (that are not completely satisfied in the Fork/Join model): the n components are independent, identical (with *exponentially* distributed Interarrival times of failures with the same mean), non-repairable, and no interference is possible between consecutive events (no queues of events are possible for the same component). The events, i.e., the *failures*, can be regarded as generated by n *independent* Poisson streams with the same mean. Denoting with $MTTF$ the mean time to *failure* of a single component, and with $MTTF_n$ the mean time to *failure* of all the n components, (its derivation is summarized in Appendix A.3) it will be:

$$MTTF_n = \left(\frac{1}{n} + \frac{1}{n-1} + \ldots + \frac{1}{2} + 1 \right) MTTF \qquad (5.3)$$

The $MTTF$ of a component represents the mean Response time R of a queue station of our model, whose values are *exponentially* distributed since each station is modeled as a M/M/1 queue. The $MTTF_n$ represents the mean time required to have the executions of all the n parallel tasks completed, i.e., the Fork/Join Response time.

Unfortunately, our original model (Fork/Join) violate several assumptions of the reliability model: the events on the four queue stations are *not* independent (the Fork generates the n parallel tasks of a job *simultaneously*), the tasks may be queued at a station to wait until the server is idle, and a task of a job may start its execution on a station *also* if the tasks of a previous job are still in execution on the other stations. However, *in spite of these violations*, the values given by Eq. 5.3 are *not very far* from the results of the simulation.

To verify these results, consider the Fork/Join Response times shown in Fig. 5.7. With $\lambda = 1$ j/s the result of simulation is $R_{FJ} = 1.922$ s while Eq. 5.3 gives 2.08 s. With $\lambda = 1.4$ j/s the R_{FJ} is 3.096 s and the approximated value is 3.47 s. With $\lambda = 1.8$ j/s the simulation provides $R_{FJ} = 9.036$ s and the approximated value is 10.4 s. If we consider a very low arrival rate, e.g., $\lambda = 0.1$ j/s the utilization is 0.05, the queues are very unlike and the simulation provides $R_{FJ} = 1.086$ s while

Eq. 5.3 gives 1.096 s, very close! Clearly, the errors increase with the queue lengths, i.e., with the arrival rate, and then with the `Utilization` of the stations.

5.1.4 Limitations and Improvements

- *Servers with different mean* `Service times`: we assumed that the `Service times` of the four `Queue` stations have the same mean and that are *exponentially* distributed. For a *generalization* it is sufficient to select `Queue` stations with different mean `Service times`, see case study Sect. 5.2.
- *Different number of tasks on each output link of a* `Fork`: the number of tasks generated by a job on each output link is the same. Generalizations are easy to implement by selecting the `Advanced Forking Strategies`, see Fig. 5.2.

5.2 Impact of Variance on Synchronization

tags: open, single class, Source/Fork/Queue/Join/Sink, JSIMg.

5.2.1 Problem Description

As in the previous problem, we consider the parallel executions of four tasks and their synchronization. The layout of the model is shown in Fig. 5.9. The only difference with respect to the problem of Fig. 5.1 stem in the variance of `Service times` of *one* of the four `queue` stations: `Queue1` has a higher variance with respect to the other three stations (all the mean values are always $S_i = 0.5$ s, as in the previous model). We want to investigate on the impact on the synchronization time of the four executions of this high-variance station.

The high variability of `Service times` is typical of many current computing infrastructures since frequently the applications are executed by very different systems. For example, the Virtual Machines that are dynamically allocated to applications have different computational power and their workloads are often unbalanced. What is surprising is that even a relatively *small difference* in the variance of the `Service times` of *one* station out of *four* (that have the same mean) has a *deep impact* on the `Fork/Join Response time`.

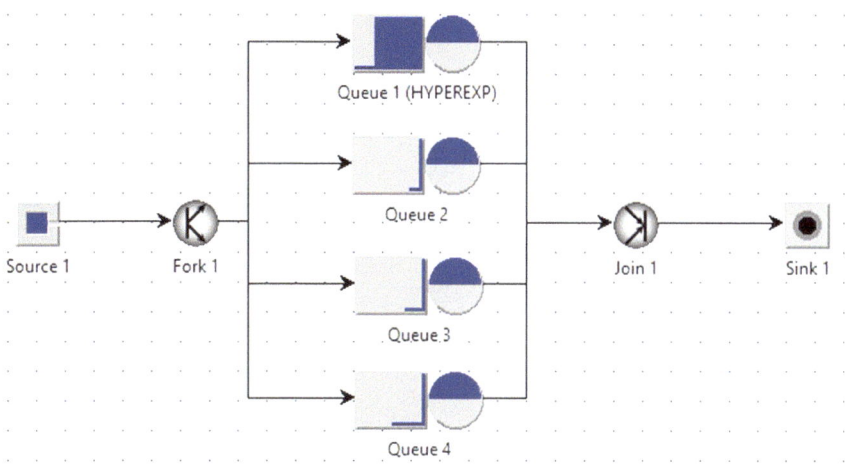

Fig. 5.9 Fork1 generates for each job four tasks executed in parallel and synchronized on Join1. Service times of Queue1 are *hyper-exp* while those of the other three stations are *exponential*

5.2.2 Model Implementation

The mean Service times of the four queue stations are the same used in the model of Fig. 5.1, $S_1 = S_2 = S_3 = S_4 = 0.5$ s. In this model, we assume that the coefficient of variation cv (given by *standard dev./mean*) of Queue1 Service times is cv = 3 instead of 1 (as it was in the previous model where we assumed exponential distributions). Thus, the standard deviation of the Service times is 1.5 s, and the variance is 2.25 s^2. Since it is cv > 1, to simulate the Service times of Queue1 we use the Hyperexponential distribution (Fig. 5.10) with parameters $cv = 3$ and mean value $S_{Q1} = 0.5$ s (see Fig. 5.10). From these two parameters JSIMg automatically derives the other parameters needed to generate an *hyper-exponential* distribution with a given mean and variance.

Initially a model with Arrival rate $\lambda = 1$ j/s is executed. The impact of the variability of Service times of one of the stations (Queue1) on the Synchronization time of the four tasks is then investigated. Comparisons with the performance of a single station M/G/1 are also done.

5.2.3 Results

A single simulation run is executed with $\lambda = 1$ j/s. The Response times of Queue1 and Join1 stations are shown in Fig. 5.11. The latter represents the Synchronization time of the executions of the four tasks.

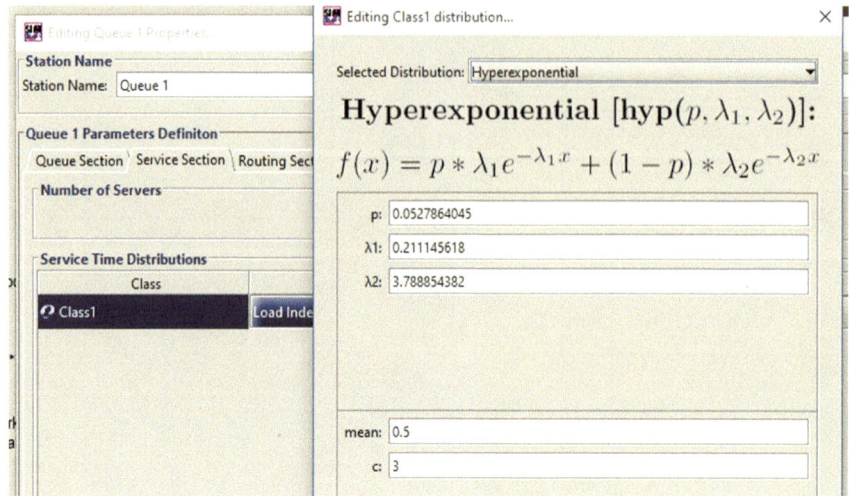

Fig. 5.10 *Hyper-exp* Service time distribution of Queue1, with mean = 0.5 and cv = 3

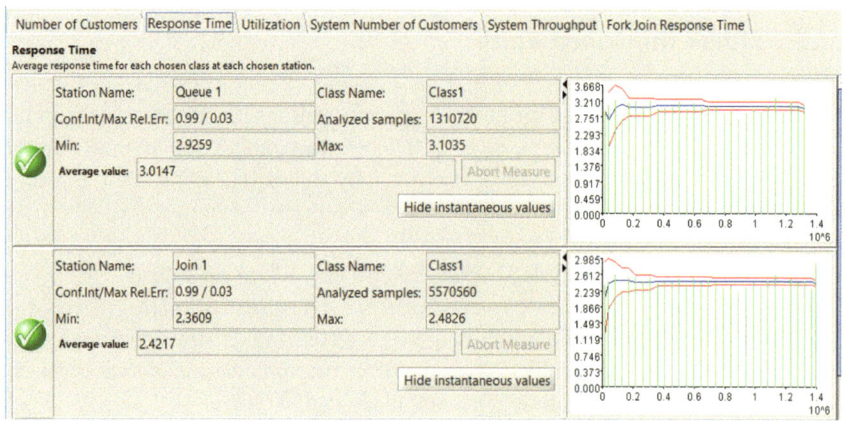

Fig. 5.11 Response times of Queue1 and Join1 stations of the model of Fig. 5.9 with $\lambda = 1$ j/s . The latter represents the Synchronization time of the four tasks

We evaluated the behavior of the Fork/Join Response time for different values of Arrival rate λ using a What-If (Fig. 5.5). Five models with $\lambda = 1, 1.2, 1.4, 1.6, 1.8$ j/s have been executed and the corresponding Fork/Join Response times are shown in Fig. 5.12.

In Table 5.2 we report for comparison purposes the Fork/Join Response times and the Synchronization times of the two models of Fig. 5.1 (column Exp) and Fig. 5.9 (column Hyper), respectively, obtained with JSIMg. The impact of the variability of Service times of one station to the Global Response time of the Fork is evident. For example, with stations utilized at

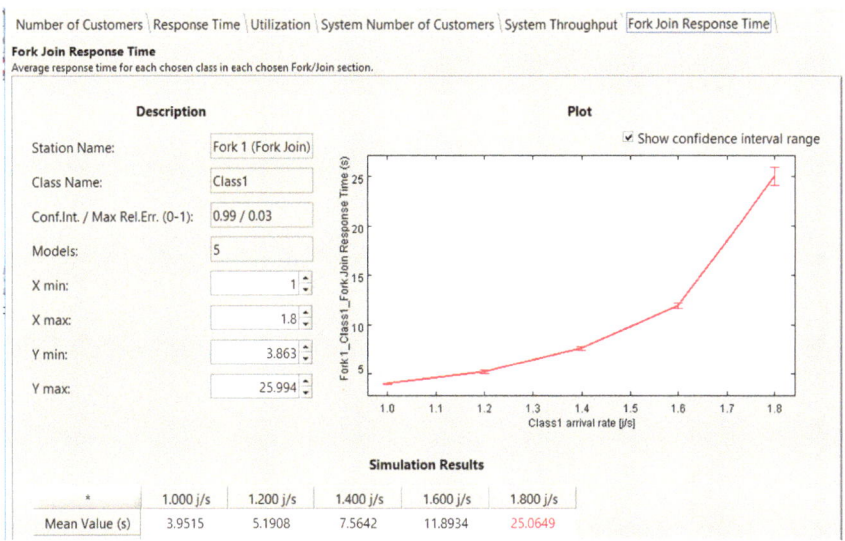

Fig. 5.12 `Fork/Join Response times` of the model of Fig. 5.9 computed with a `What-if` with the `Arrival rate` λ ranging from 1 to 1.8 j/s in 5 steps (Fig. 5.5)

90% the `Fork/Join Response time` increases from 9 s (when the standard deviation of `Service times` is 0.5 s) to 25 s (when the standard deviation of is 1.5 s).

Let us remark that it has been sufficient that *only one* of the four servers increased its variance of `Service times` of three times to generate a similar increase of the `Response Time` of the `Global Fork/Join` structure.

To analyze the impact of the `Queue1` station with the high variance on the `Fork/Join Response time` we study it in isolation. According to the assumptions, the `Interarrival times` of the tasks are *exponentially* distributed and its `Service times` follow an *hyper-exponential* distribution. Thus, `Queue1` can be modeled analytically as a M/G/1 queue. Its `Response Time` is given by (see, e.g., [36]):

$$R_{Q1} = S + \frac{US(1 + cv^2)}{2(1 - U)} \tag{5.4}$$

where U is the `Utilization` of the station (U = λ S), S = 0.5 s is the mean of `Service times` and cv = 3 is their coefficient of variation. In the last two columns of Table 5.2 the `Response Times` of two stations M/G/1 and M/M/1, considered in isolation, are reported. The contribution of the M/G/1 station to the global `Fork/Join` performance is evident if we consider, for example, that with λ = 1.8 j/s its `Response time` (23 s) represents the 92% of the `Fork/Join Response time` (25 s) with the four `queue` stations. It must also be pointed out the huge difference between the `Response Times` of the `two` types of queues

Table 5.2 Fork/Join Response times and Synchronization times of the two models of Figs. 5.1 (label Exp, four exp) and 5.9 (label Hyper, one *hyper-exp* and three *exp*)

Arrival rate	Util.	Fork/Join Response time		Synchronization time		Resp.time single station	
[j/s]	%	**Hyper**	Exp	**Hyper**	Exp	M/G/1	M/M/1
$\lambda = 1.0$	0.5	**3.951**	1.922	**2.463**	0.920	3.00	1
$\lambda = 1.2$	0.6	**5.191**	2.338	**3.197**	1.103	4.25	1.25
$\lambda = 1.4$	0.7	**7.564**	3.096	**4.680**	1.443	6.333	1.666
$\lambda = 1.6$	0.8	**11.893**	4.510	**7.244**	2.046	10.50	2.5
$\lambda = 1.8$	0.9	**25.065**	9.036	**15.875**	4.024	**23.00**	**5**

M/G/1 and M/M/1, for example, with U = 0.9 the two values are 23 and 5 s, respectively (last two columns of Table 5.2).

5.2.4 Limitations and Improvements

- *High variability of* Service times: All the servers considered in the models have the same mean. It is easy to *generalize* these models considering *heterogeneous* servers with different *mean* Service times and *distributions*.
- *Impact of high variability of* Service times *of one server*: Let us remark that it has been enough an increase of the variance of Service times of *only one* server out of *four* to generate dramatic effects on system performance. You may imagine how frequently this condition occurs in real world data centers with the high degree of heterogeneity of current workloads! It is therefore very important to keep the variability of Service times of *all* the servers under control.

5.3 Synchronization on the Fastest Task

tags: open, single class, Source/Fork/Queue/Join/Sink, JSIMg.

5.3.1 Problem Description

In this section we will analyze the effects on the Fork/Join Response time of a Join Strategy different from the Standard one (that synchronizes the executions of *all* the tasks). According to the Quorum strategy, a Join station releases a job, i.e., fire the job, when a *subset* of the parallel tasks generated by the Fork for each job completed their execution. In this problem we assume that as soon as one task of a job completed its execution, the Join releases the job.

This problem is typical of several actual digital infrastructures like CEPH, the object storage used by *OpenStack*, or RAID1, the mirroring storage architecture, that use data *replication* as a technique to improve performance and reliability of systems. The requests for a object (data, file or other subject) are split in several tasks that are sent in parallel to all the devices containing the replicated data. In our case, the object is sent back when the first task (the fastest) finishes. The results show that the impact of replication technique to the performance and reliability of digital infrastructures is *significant*.

5.3.2 Model Implementation

We consider the parallel executions of four tasks generated by a job at the `Fork` on four `Queue` stations having the *same* characteristics.

The service requests of the four tasks have the *same* mean $S_1 = S_2 = S_3 = S_4 = 0.5$ s and are exponentially distributed. The arrival rate of the jobs is $\lambda = 1$ j/s, and the interarrival times are exponentially distributed. The layout of the model is shown in Fig. 5.13. The difference of this model with respect to the one considered in Sect. 5.1 is that the `Join` do not requires that *all* the four executions must be completed before releasing the job but it is sufficient that *only one* of them (i.e., the fastest) completes. We will use the `Join Strategy` with Quorum=1 (see Fig. 5.14).

5.3.3 Results

The behavior of performance indexes for different values of arrival rate λ is investigated using the `What-if` (see, e.g., Fig. 5.5). The `Arrival rate` is selected as control parameter and the solution of *seven* models is requested with $\lambda = 0.25, 0.5, 0.75, 1.0, 1.25, 1.5, 1.75$ j/s, respectively.

The mean `Fork/Join Response times` are shown in Fig. 5.15. We want to emphasize the differences between the mean values of this index obtained with Quorum=1 (the `Join` releases the job when the shortest task completes its execution) and the ones obtained with Quorum=4 (the `Join` wait that *all* the executions of *four* tasks are completed before release the job). Table 5.3 shows the values of these indexes in the first two columns for the `Arrival rates` λ ranging from 0.25 to 1.75 j/s.

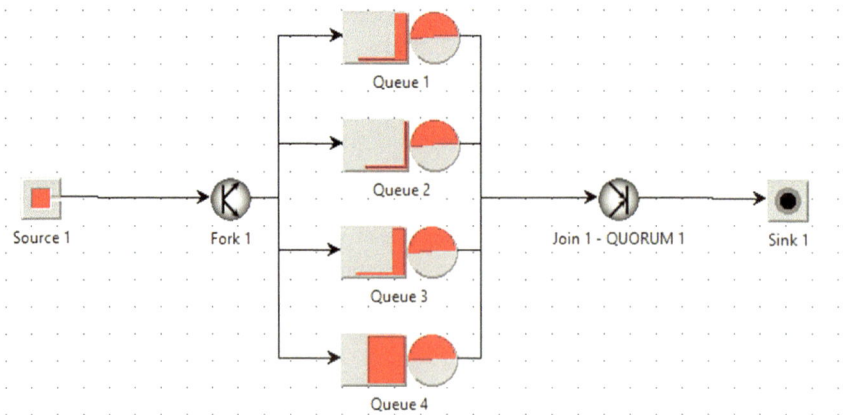

Fig. 5.13 `Fork1` generates four *tasks* executed in parallel for a job: `Join1` waits only the *fastest*

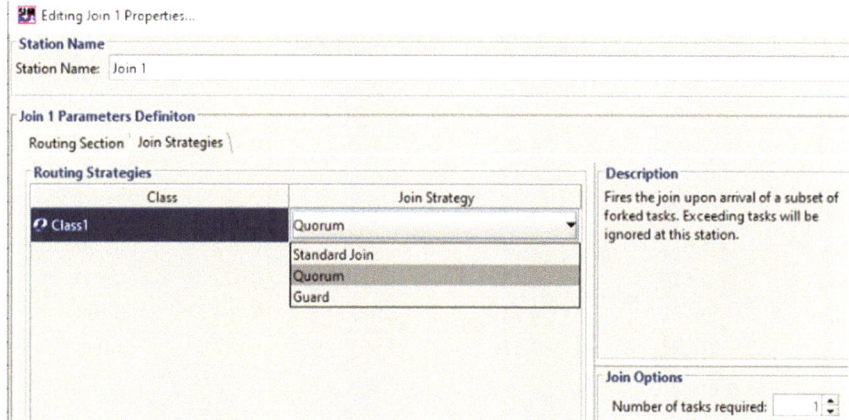

Fig. 5.14 Selection of the `Quorum Strategy` of the `Join`

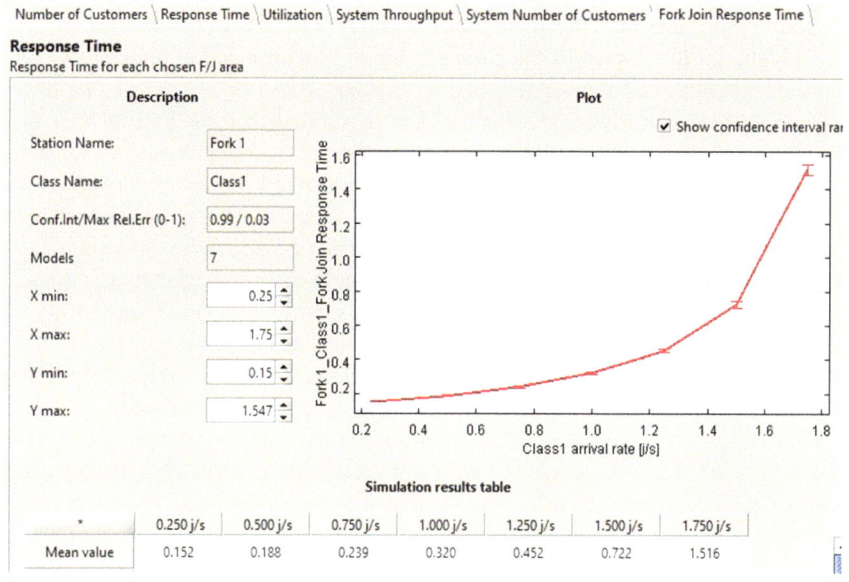

Fig. 5.15 `Fork/Join Response times` with `Quorum=1 Strategy` obtained with a `What-if`

Table 5.3 Fork/Join Response times with Join Strategies Quorum=4 (Join waits all four tasks) and Quorum=1 (Join waits only the fastest task)

Arrival rate	Fork/Join Response time		Queue1	Queue1	Optim.
λ [j/s]	Quorum = 4	**Quorum = 1**	Resp.Time	Utiliz.	Approx.
0.25	1.165	**0.152**	0.570	0.125	0.142
0.5	1.328	**0.188**	0.670	0.248	0.166
0.75	1.558	**0.239**	0.793	0.376	0.200
1.0	1.915	**0.320**	1.006	0.501	0.250
1.25	2.513	**0.452**	1.321	0.625	0.333
1.5	3.722	**0.722**	2.013	0.764	0.500
1.75	7.227	**1.516**	3.963	0.877	1.000

As can be seen, the differences between the values obtained with Quorum=1 and Quorum=4 are remarkable. For example, for $\lambda = 1$ j/s the value with Quorum=1 is about *6 times less* than the value with Quorum=4 (0.320 s vs. 1.915 s)!

To highlight the impact of the parallelism and synchronization policies, we also show in the Table 5.3 the Response time and Utilization of one of the Queue stations considered in isolation. Let us remind that all the four queues are the same. Since with Quorum=1 the Fork/Join Response time is the Response Time of only one task, it may seem correct to consider only one of the queue in isolation to compute its value (see the Queue1 Resp.time column). This assumption is wrong. Indeed, the Response times with Quorum=1 are considerably lower than those obtained with a single queue station (e.g., with $\lambda = 1$ j/s it is 0.320 s vs. 1.006 s!). The error occurs because with Quorum=1 only the *minimum* of four sequences of exponentially distributed execution times (with the same mean) is considered, while with the single queue only the average of a single sequence of exponentially distributed Service times is considered.

An *estimate* very easy to compute (referred to as Optimal Approximation) of the Fork/Join Response time with Quorum=1 can be obtained considering the end of each task as *events* generated by *n* independent poissonian generators with the same rate $1/R$. This modeling approach is often used to study the reliability of a system consisting of *n* components, e.g., devices, connected in series. This type of systems fails when any one of the *n* components fails. The events considered are the *failures* of devices. The time between two consecutive failures of a device is referred to as MTTF, *mean time to failure*. The assumptions considered are: *independence* of the *n* identical components, the failures are *exponentially* distributed in time, and the components are *non-repairable*. Thus, we may consider the model as consisting of *n* identical independent poissonian arrival streams of events (the failures) with interarrival times exponentially distributed, and *no queues* are possible

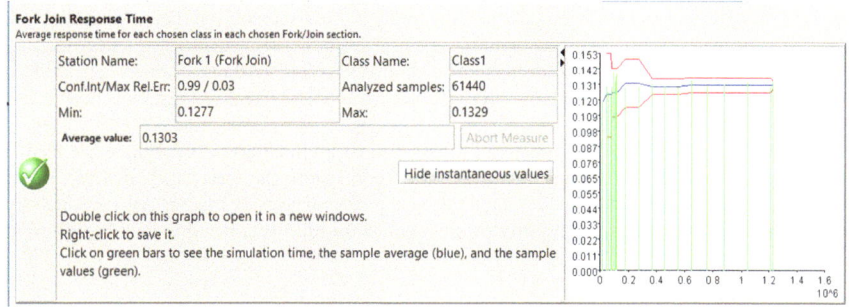

Fork Join Response Time
Average response time for each chosen class in each chosen Fork/Join section.

Station Name:	Fork 1 (Fork Join)	Class Name:	Class1
Conf.Int/Max Rel.Err:	0.99 / 0.03	Analyzed samples:	61440
Min:	0.1277	Max:	0.1329
Average value:	0.1303		Abort Measure

Hide instantaneous values

Double click on this graph to open it in a new windows.
Right-click to save it.
Click on green bars to see the simulation time, the sample average (blue), and the sample values (green).

Fig. 5.16 Fork/Join Response time with Join Strategy Quorum=1 and $\lambda = 0.05$ j/s

among consecutive events. The mean time $MTTF_1(n)$ for the *first* failure of such a systems (see Appendix A.3 for its derivation) is given by:

$$MTTF_1(n) = \frac{MTTF}{n} = \frac{R}{n} \qquad (5.5)$$

where *MTTF* represents in our model the mean Response time R of a queue station that is exponentially distributed. Indeed, according to the assumptions, each queue is of M/M/1 type, and thus it is $R = S/(1 - \lambda S)$ with exponential distribution. Considering that the number n of parallel stations is 4, Eq. 5.5 provides the values reported in the last column Optim.Approx. of Table 5.3.

As can be seen, these values are not very far from the corresponding Fork/Join Response times with Quorum=1 and the differences increase with λ. This is due to the assumptions made in the failures model that are violated in the simulated model of Fork/Join. Indeed, the parallel tasks of a job are generated by the Fork simultaneously on the n queue stations, so are *not independent*, and furthermore *interferences* are possible among consecutive tasks at any of the n queue.

If we consider a *very low* Arrival rate, e.g., $\lambda = 0.05$ j/s, the Fork/Join Response time given by Eq. 5.5 is 0.128 s and the value obtained with JSIMg is 0.130 s (Fig. 5.16). These values are so close because in this case the Utilization of the queues is very low, $U = \lambda S = 0.025$, and thus the interferences among consecutive tasks are *negligible*. Indeed, the Response time of a Queue station given by the simulation is 0.514 s and the mean Service time S is 0.5 s, very close (practically, queues of tasks waiting for the server almost never form). Clearly, with the increase in the arrival rate, the approximation becomes increasingly losing.

Chapter 6
Reference Models

6.1 A Facial Recognition Surveillance System

tags: open, two classes, Source/Queue/Class-Switch/Sink, JSIMg.

We consider a surveillance system consisting in the facial identification of passengers flowing in an airport. It is implemented with a Edge computing architecture. Similar systems can be applied in several environments such as railway stations, shopping malls, roads, airways, banks, public buildings, museums, hospitals, etc. It is a simple model that represents a first step towards the solution of the complex problems of security control.

6.1.1 Problem Description

Currently available *Internet of Things* (IoT) devices are equipped with powerful processors, large storage and actuators that generate huge amounts of data that must be transmitted through the network. Cloud computing, with its characteristics of large availability of highly scalable servers, is very appropriate also for the IoT-based architectures. However, since the distance between the IoT devices and the cloud servers is typically large, the resulting latency is *not negligible* and exhibit unpredictable fluctuations.

  This characteristic is *very negative* for most IoT applications that are *delay-sensitive* because are based on *decision/reaction* cycles (see, e.g., virtual reality, smart building, video surveillance, facial recognition, e-health, monitoring, automotive and traffic control). *Minimizing* the time required to process the data generated by the IoT devices is essential for the correct execution of these applications.

  To approach this problem, massively distributed architectures that allow the implementation of the *Edge computing* paradigm have been introduced. In these

© The Author(s) 2024
G. Serazzi, *Performance Engineering*,
https://doi.org/10.1007/978-3-031-36763-2_6

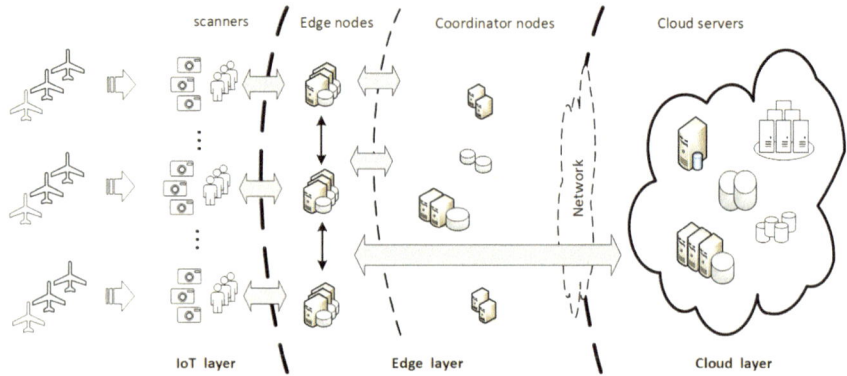

Fig. 6.1 The surveillance system based on the identification of facial scans of passengers

environments, the components, referred to as Edge nodes, that process the data
are placed *as close as possible* to IoT devices, i.e., at the *edge* of the network.

Typically, the Edge nodes have sufficient processing power and storage capacity to execute *efficiently* most of the tasks of the applications. Coordinator servers
that perform application management tasks (when they are needed) are placed near
the Edge nodes and are connected to them with fast links. Only the *heaviest tasks*
are sent to the Cloud servers.

The latency reduction is achieved in two ways: on one side most of the tasks
are executed locally by the Edge nodes in close proximity of IoT devices, on
the other side *only* the heaviest tasks are sent to powerful cloud servers. As
a consequence, the data transmitted over the network and the Response times
can be *minimized*.

In this case study we describe the *Edge computing* environment (see Fig. 6.1)
which is used in an airport to implement a surveillance system based on facial recognition. The identification system detects the faces of passengers passing by the
scanners, those that go up and down the escalators, those in line at check-in desks,
and those flowing in various areas of the airport (e.g., waiting rooms, shops, bar,
restaurants). Five *categories* of persons are considered, corresponding to five types
of scans: *poor-quality image, regular, suspect, dangerous,* and *unknown* person.

To identify the category to which they belong, the faces detected by a scanner are
first compared with those of an in-memory database stored in the directly connected
Edge node. The *reaction actions* that must be taken after a scan identification vary
greatly depending on its category. The scans belonging to *poor-quality* and *regular*
(safe people) categories only require accesses to the local databases but no further
actions. The scans of *suspect* and *dangerous* categories require, among others, *very
quick* actions to synchronize the scanners along the path followed by the person to
be tracked and must transmit messages to the interconnected security agencies.

The algorithms for matching the detected faces with the images stored in the local
in-memory database are executed on the Edge nodes. The Edge nodes also

interact with each other to synchronize the in-memory databases and to coordinate reaction actions.

The scans of the *unknown* category (i.e., those not present in the local `Edge nodes` databases) are sent to the *cloud* for more in-depth analysis. At the `cloud layer`, very large NoSQL distributed databases for Big Data (such as Apache HBase, Hive, Cassandra, Mongo DB), with documents, social media profiles, biometric data, and voice traces are used for extensive identification analysis with the most advanced face detection algorithms. Machine learning algorithms are implemented to train the system to minimize false identifications. The results of these additional processing are sent back to the `Edge nodes` to update the local databases and then sent to the `System Coordinators` to implement the *reaction actions*. To make the presentation simple, in the implemented model we have *not explicitly considered* the `System Coordinator` servers as very often they are not present or are powerful servers that do not cause performance problems.

The capacity planning study is structured in *two main phases*, referred to as *initial sizing phase* and *performance forecast phase*, respectively.

The objective of the *initial sizing phase* is the calculation of the number of `Edge nodes` that guarantee the achievement of the performance targets with the planned workload (referred to as *original workload*). The most important performance constraint is the time required to analyze a scan by `Edge nodes`, i.e., their mean `Response time`, that must be *less than 3 s* for all scan categories, *excluding* the *unknown*. This constraint is important as most of the *reaction actions* to be effective must be activated within 3 s from the image detection. The configuration of each `Edge node` consists initially of one server mounted in a rack located in a dedicated room. The computed number of `Edge nodes` coincide with the number of server rooms. To ensure the highest level of physical safety of the equipments, the locations of the rooms are kept secret. To increase the *availability* of the global system, each room has independent equipments for fire detection, flooding protection, cooling and power continuity UPS. For several important reasons, the number of server rooms *can not change* over a long period of time. Initially, the scan flows arriving at the `Edge nodes` are considered *balanced* across all nodes. This first phase of the study is also important for its connections with the building constructions of the airport.

The *second phase* of the study is devoted to the assessment of the impact of different *workload growth* patterns on the `Response times` of `Edge nodes`. Several factors, in fact, as the results of the commercial policies of the airlines or the success of the new destinations served, make the forecast of workload trends very uncertain. In short periods of time, huge differences can occur between the scan streams arriving at the various `Edge Nodes`. Therefore, it is required that the implemented model be able to simulate very different workloads in terms of arrival patterns and mix of scan categories. *Two types* of workload growth are considered: increase of traffic *intensity* keeping fixed the fractions of scan categories of the original workload, and workloads with significant differences in the *mixes* of categories in execution. The impact of these types of workload changes on the performance of the global system are studied. This knowledge is fundamental for the implementation of the *scalability feature* of the `Edge nodes` with respect to the workload growth.

The main results of this case study are:

- computation of the number of Edge nodes required to meet the *performance target* of their *Response times* with the original workload
- identification of the number of scanners for each Edge node in order to provide a balanced load of scans based on the airport map on the Edge nodes as a function of the characteristics of the layout of the airport and differences (in intensity and mix of scan categories) of passenger flows. This result allows to identify the physical locations of Edge nodes in the various buildings
- computation of the number of servers for each Edge node required to meet their performance target as a function of the workload growth. The computed performance metric can be used to drive the *horizontal scaling* component (that can be implemented in the system) of each Edge node *separately*
- show how a complex model can be *decomposed* into several simpler models that can be solved separately, the results thus obtained can be combined to provide the solution of the global model (see, e.g., the *incremental approach* in Sect. 1.1 and Fig. 1.2).

6.1.2   Model Implementation

The *scanners* generate the *face scans* sent in input to the model and represent the Source of the identification requests arriving at the Edge nodes (Fig. 6.2). Depending on the processing time and the path between the resources, the requests can be divided into two groups, i.e., *two classes*. The first class (class-E) comprises the scans belonging to the *poor-quality, regular, suspects,* and *dangerous* categories, that are *completely* processed by the Edge nodes. The second class (class-C) consists of the scans of the *unknown* category that require additional processing by the Cloud servers.

Figure 6.2 shows the model of the global system. The solid lines represent the path between the resources of class-E requests while the dashed lines represent the path of class-C requests.

All the requests sent by the Source stations to the Edge nodes are *initially* of class-E type. When their identification process is completed, requests that belong to *poor-quality, regular, suspect,* and *dangerous* categories leave the model through Sink class-E stations. They will be subsequently processed by the System Coordinators servers, not considered in the model presented here, to implement the *reaction actions*. The requests of the *unknown* category are instead routed to the Cloud servers for a more extensive analysis. The class of these requests is *changed* from class-E to class-C in the Class-Switch station before joining the Cloud servers. Once their processing is completed, these class-C requests are sent back to the Edge nodes to update the local in-memory databases and other data structures before leaving the model through the Sink class-C stations. In Fig. 6.2, p_c represents the fraction of the requests generated by the

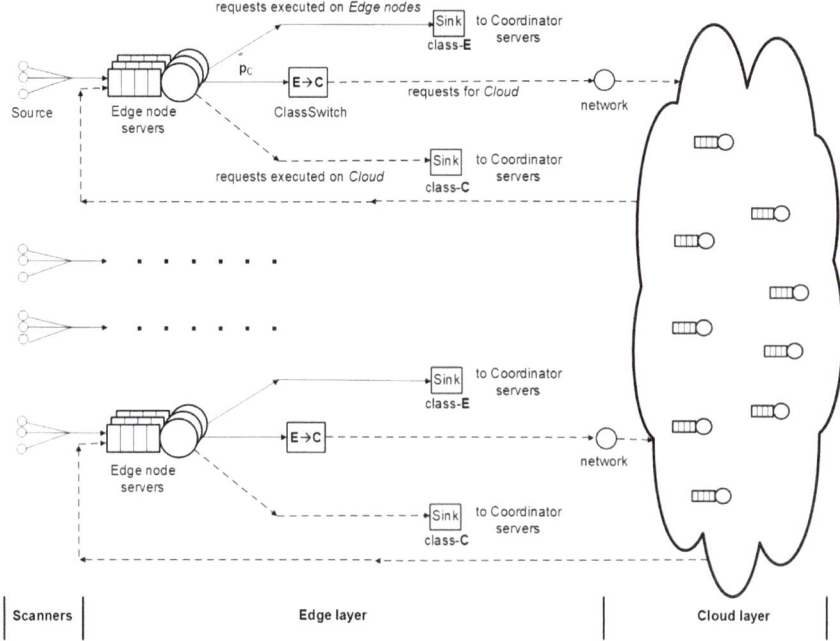

Fig. 6.2 Model of the global facial recognition system. Solid lines represent the path of `class-E` requests while the dashed lines the one of `class-C` requests

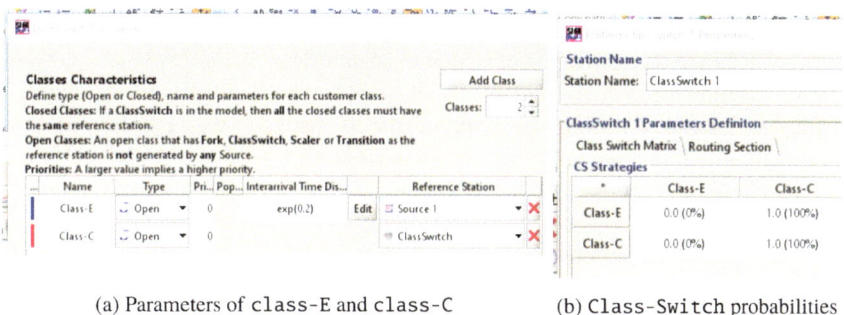

(a) Parameters of `class-E` and `class-C` (b) `Class-Switch` probabilities

Fig. 6.3 Parameter settings of `class-E` and `class-C` requests (**a**); probability that a request change class in the `Class-Switch` station (**b**)

scanners connected to an `Edge node` that is sent to the `Cloud servers`, i.e., those that belong to the *unknown* category.

An example of the parameter settings of the two classes is shown in Fig. 6.3a. Note that only `class-E` requests are generated by the `Source` (in the figure, their arrival rate is set to $\lambda = 0.2$ req/s with exponential distribution) because the `class-C` requests are generated in the `Class-Switch` station from the arriving `class-E` requests (thus no parameters must be set). As a consequence, the `Reference`

Table 6.1 Service demands [s] of class-E and class-C requests. p_c is the fraction of class-E requests that belong to *unknown* category, sent to Cloud servers as class-C requests

Resource	Class of requests	
(Station)	E	C
Edge node	0.5	0.1 p_c
Cloud server	–	0.8 p_c

stations of the two classes are Source and Class-Switch, respectively. The selection of different Reference stations is important for the computation of the correct values of the per-class performance indexes.

The window for the definition of the parameters of the Class-Switch station is shown in Fig. 6.3b. In the considered problem it consists of a 2x2 matrix, whose entry i-j represents the probability that a class-i request entering the station will be changed to class-j when it exit. In our model this matrix is simple as the arriving requests at Class-Switch station are only of class-E and are all changed to class-C. Indeed, the class-C requests arriving at the Edge node after being processed by the Cloud servers, are sent directly to the Sink class-C station by the *routing algorithm.*

For each Edge node an instance of a Virtual Machine (VM) is launched in a Cloud server to process *all* the requests sent by that node.

The global processing time required by the face recognition algorithm to solve the pair matching problem (i.e., to find which person among the set of the local in-memory database the scan represents, if any) on an Edge node is $D_{E,E} = 500$ ms. The time required by a scanner to detect, pre-process and transmit an image is negligible compared to the time required for its analysis. We take care of it by applying a small increase in the service demand $D_{E,E}$.

Scans of category unknown are sent to the Cloud servers as class-C requests, and require $D_{C,C} = 800$ ms for their processing. The results of this analysis are sent back (still as class-C requests) from Cloud servers to Edge nodes that require additional $D_{E,C} = 100$ ms for their analysis (to update several data structures and the local in-memory database) before sending them to Sink class-C. The Network is not represented in the model as a separate component since the transmission time of data to and from the Cloud servers is negligible with respect to their processing demands.

Table 6.1 summarizes the mean values of the Service demands(exponentially distributed) of the two classes of requests. The weight p_c of the class-C demands is introduced to take into account that only the fraction p_c of the requests generated by the scanners is sent to the Cloud servers as class-C requests. The difference between the processing time of the two class of requests is very large: class-C require 900 ms while class-E require 500 ms! Clearly, the value of p_c *deeply* influences the performance of the system. Therefore, by changing the value of p_c from 0 to 1, we can model all the possible configurations of the workload.

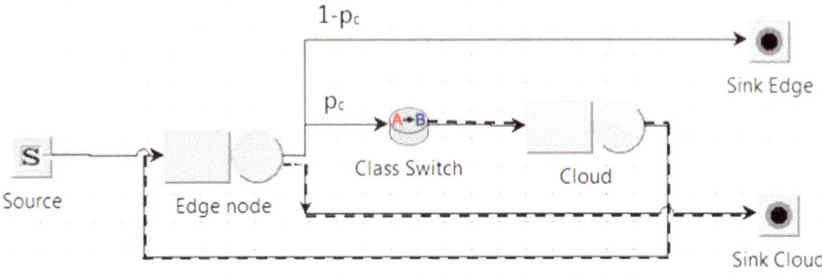

Fig. 6.4 JSIMg model of an *elementary component* consisting of *one* Edge node. Solid lines represent the path of class-E requests while the dashed lines the one of class-C requests

Important *simplifications* of the global system model can be obtained by applying the assumptions introduced in the project description. These made it possible to adopt the incremental approach technique (see Fig. 1.2). Among them, the most important are: once computed in the initial phase, the number of Edge nodes must be kept *constant* while the number of servers of each node can *increase* as a function of the performance requirements, the scanners *cannot change* the Edge node to which they are directly connected but their number can change according to several parameters (e.g., high or low traffic locations serverd, bursts of arriving people, layout of the building), the fraction p_c of the *unknown* scans received by the Edge nodes is initially considered the same for all nodes, the load of each Edge node can vary according to several parameters that are *dependent* only from the node itself, there is no interference among the VM instantiated in the cloud by the various nodes.

As a result of these assumptions, Edge nodes can be considered *independent* from each other and the global system model can be *subdivided* into as many simple models, referred to as *elementary components*, as there are Edge nodes. The model of an *elementary component* is shown in Fig. 6.4. Therefore, we can approach the capacity planning problem of the overall system by investigating the performance behavior of each Edge node separately.

As requested by the application, the mean time required by the analysis of a scan by an Edge node *must be* less than 3 s, i.e., for class-E requests it *must be* $R_{Edge}^E \leq 3$ s. Each Edge node must meet this performance constraint processing scans of *all* categories except for the *unknown* ones.

In the *initial sizing phase*, the overall intensity of the original workload has been subdivided *evenly* among all the nodes. The load of each node is assumed the same both in *intensity* and *composition* (i.e., *mix* of scan categories) for all the nodes. The number of Edge nodes computed in this phase, initially configured with one server each, is the *minimum* required to satisfy the performance constraint with the original workload. The correspondent arrival rates, referred to as *guard values*, are considered as thresholds that cannot be overcome. When a node guard value is reached (or rather approached), a new server will be allocated on its rack (or switched to on-line status if it is already mounted) and the incoming requests to the node will be balanced among all servers in the rack. This scaling policy is applied on all `Edge nodes` separately.

The `JSIMg` tool has been used to implement the simulation models.

6.1.3   Results

The objectives of the study were many. In what follows we will describe the activities regarding the following two:

— **Obj.1**: *Initial sizing and Dynamic Scalability of the* `Edge nodes`
— **Obj.2**: *Investigate the behavior of the* `Response times` *of the* `Edge nodes` *as a function of the mix of scan categories in execution*

— Obj.1: Initial sizing and Dynamic Scalability of the Edge nodes.
In the design-phase of the project, the arrival rate of scans for the entire airport, referred to as *original workload*, is set to $\lambda_0 = 42$ scan/s. The fraction of the detected scans that belong to the *unknown* category is 40% ($p_c = 0.4$) and is assumed to be the same for *all* `Edge nodes`. The service demands of the two classes of requests, are shown in Table 6.1. According to the hypotheses, the *original workload* of rate λ_0 is subdivided *evenly* among all the nodes N_{EN}. We considered the model of an individual `Edge node` (see Fig. 6.4) subject to arrival rates of scans ranging from 0.2 to 1.75 scan/s. Let us remark that the saturation load of an `Edge node` is $\lambda^{sat} = 1/[0.5+(0.1 \times 0.4)] = 1.85$ scan/s. The `Response times` of an `Edge node` in the initial configuration with one server, obtained with a `What-if` analysis, are shown in Fig. 6.5. We assume *exponential* `Interarrival times` of the scans, and PS (Processor Sharing) scheduling policy in the queue stations that is typically used to simulate multiclass workload parametrized with the service demands. This policy capture the reality better than FCFS since the `Service demands` are obtained summing the `Service times` of all the visits of a request. So, with PS, the executions of all the requests are seen to progress concurrently also at the demands level. Furthermore, several analytical solvers (see JMVA) when this policy is adopted provide exact solutions of models with multiclass workload (see, the BCMP theorem in Sect. 3.1).

The threshold value of the average processing time (i.e., the `Response time`) of `class-E` scans of the `Edge nodes` is 3 s. According to the hypothesis that the original workload is *initially equally divided* among all the `Edge nodes`, we computed their minimum number N_{EN} needed to satisfy the constraint. With $N_{EN} = 30\ servers$, the arriving rate of scans at each node is 1.4 scan/s (λ_0/N_{EN}), and the correspondent mean `Response time` of the `Edge node` with one servers is $\simeq 2$ s (see Fig. 6.5).

Higher arrival rates, e.g., 1.55 scan/s, could also have been considered. But the adoption of $\lambda = 1.4$ scan/s as guard value for the *scalability monitor* is motivated by the tolerance of unexpected fluctuations in the flow of passengers without seriously violating the constraint of `Response times` (e.g., the 10% increase in load

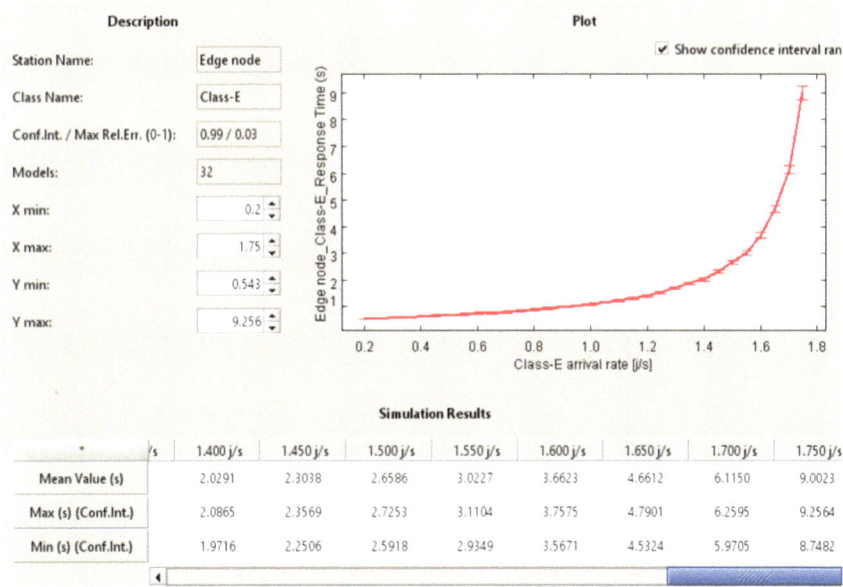

Fig. 6.5 `Response times` [s] of an `Edge node` with a single server for arrival rates of scans $\lambda = 0.2 \div 1.75$ scan/s, 40% of them belong to the *unknown* category ($p_c = 0.4$)

corresponds to a `Response time` of $\simeq 3$ s, which in any case still satisfies its limit value).

The global number of scanners N_{Scan} to be installed in the airport has been computed considering the technical characteristics of the scanners, the processing capacity of the Edge servers, and the intensity of the flow of passengers in the airport. The result of the computation is $N_{Scan} = 840$, an average of 28 scanners per node. Let us remark that due to the heterogeneity of traffic in the paths of the airport, this initial subdivision does not correspond to the assignment of the *same number* of scanners per each node but to the one that generate a *balanced* load across all nodes. With the arrival rate of 1.4 scan/s at each node, and the average of 28 scanners per node, each scanner generate an average of 1 scan every 20 s.

To avoid instability, a new server is allocated in a node when its arrival rate is higher than the guard value for a *time interval* whose duration depends on the path considered. The load will be re-balanced among all installed servers after a transient period considering the arrival of new requests and the exit of executed requests. In Fig. 6.6 the behavior of the `Response times` of an `Edge node` with arrival rate ranging from 0.2 to 10 scan/s and a number of servers from 1 to 6 is shown. This diagram is fundamental for the *horizontal scalability* of the `Edge nodes` since it provides the number of servers that are needed to meet the performance constraint on `Response times` as a function of the load behavior. When an autoscaling component is used, it provides the values for triggering scale-up actions.

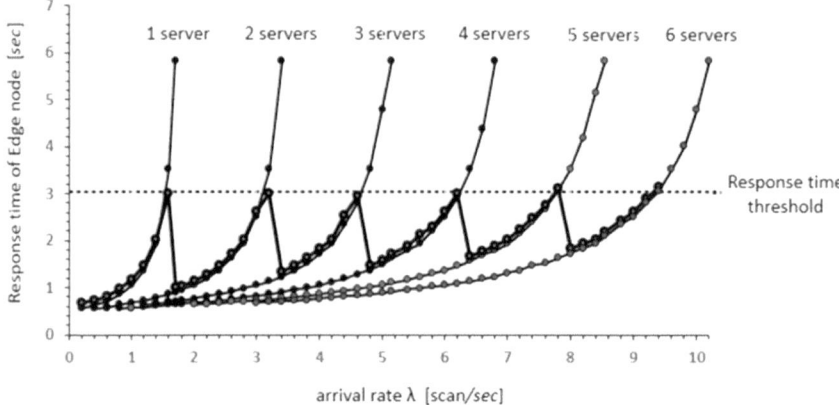

Fig. 6.6 `Response Times` of an `Edge` node for the analysis of a face-scan *versus* the arrival rate for various number of servers; 40% of the arriving scans belong to the *unknown* category ($p_c = 0.4$)

— Obj.2: Investigate the behavior of the `Response times` of the `Edge` nodes as a function of the mix of scan categories in execution.

In systems with multiclass workloads, the *bottleneck* (i.e., the resource with the highest utilization) can migrate between resources depending on the mix of classes of requests being executed. The greater the difference between the maximum service demands of the classes (when they refer to different resources) the deeper the effects of bottleneck *migration* on system performance. For example, in our system when the load consists of `class-E` scans only (i.e., with $p_c = 0$) the maximum Throughput is $X_0^{max} = 1/D_E^{max} = 2$ scan/s while with `class-C` scans only (i.e., with $p_c = 1$) it is $X_0^{max} = 1/D_C^{max} = 1.25$ scan/s (37.5% less!).

Thus, it is important to consider the resource *utilizations* of *each* class of requests. The capacity planning study must evaluate the projections on performance of all possible changes in the workload, not only in terms of intensity but also in the mix of classes of requests being executed.

In *Obj.1* the fractions of scan categories arriving at the `Edge nodes` (initially they are all of `Class-E`) were considered constant: 40% of them were of *unknown* category ($p_c = 0.4$). Now we relax this assumption and investigate the behavior of the `Utilizations` and `Response times` with all the possible mix of scan categories. By applying the utilization law $U_i = \lambda D_i$ to the `Edge` and `Cloud` servers of the open model of Fig. 6.4, we can easily obtain the mix of requests that balances the load on the two resources.

By simplifying the two equations of λ and equating the global service demands D_E and D_C (see Table 6.1) we have $0.5 + 0.1 p_c = 0.8 p_c$. Thus, we can derive the fraction $p_c = 0.71$ of all incoming scans that generate the *equiutilization* of the resources. For $p_c < 0.71$ the most utilized resource is the `Edge node`, while for $p_c > 0.71$ it is the `Cloud server`. Figure 6.7 show the behavior of `Utilizations`

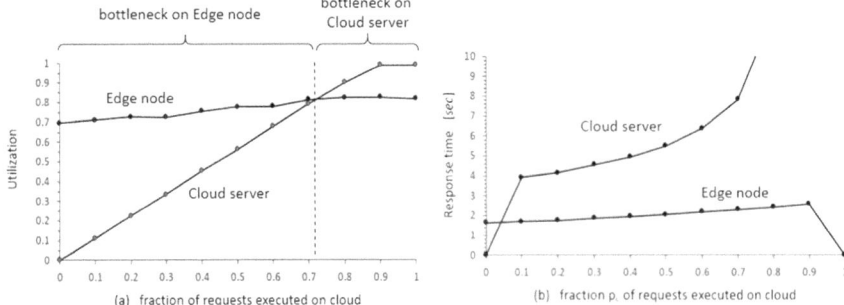

Fig. 6.7 Utilization and Response time of the Edge node and the Cloud server with respect to the mix of scan categories in execution, with arrival rate $\lambda = 1.4$ scan/s

and Response times of the Edge node (with one server) and the Cloud server for *all* the possible mix of scan categories in execution with arrival rate of scans $\lambda = 1.4$ scan/s.

The impact of the different mixes on the performance of the two classes of requests is evident. As computed above, the mix corresponding to the equiutilization of the two resources is obtained with $p_c = 0.71$. The Utilizations of Edge node range from 0.7 (with $p_c = 0.1$) to 0.8 (with $p_c = 0.9$) and the corresponding Response times are 1.7 and 2.9 s. Note that the Cloud server Response time increases rapidly when its arrival rate of class-C requests approach the saturation value of $1/D_C^{max} = 1/0.8 = 1.25$ scan/s. Since the arrival rate of class-C requests to Cloud server is $\lambda\, p_c$, its utilization is given by $U_C = \lambda\, p_c\, D_{C,C}$. The values of p_c that generate the saturation can be easily obtained from this equation considering $U_C = 1$, $\lambda = 1.4$ scan/s, and $D_{C,C} = 0.8$ S. The result is $p_c = 0.892$. Clearly, the models with higher values of p_c are not in equilibrium and are unstable since one resource is saturated. To improve the performance of the system with higher values of p_c it is necessary to use more powerful VMs of the Cloud servers.

6.1.4 Limitations and Improvements

The model described is clearly a simplified version of a global surveillance system model. However, with limited effort it can be improved in different directions. Among them are:

- *Fractions of scan categories*: The assumption that the fraction of scans of *unknown* category is the same for *all* the Edge nodes is a limitation that can be easily relaxed. In this case, it is enough to make a model for each distinct Edge node with the fractions of scan categories arriving at the node. In many cases it is sufficient to identify groups of Edge nodes having similar characteristics with respect to the flow of passengers and the fraction of scan categories and to implement only their models.

- *Interarrival time distributions of scans*: To capture the differences of arriving traffic of scans among the various categories several classes can be considered. So, for example, bursts can affect one category while another one can have a different distribution. For each class, follow the sequence `Define customer classes`, `Edit`, and select the distribution from the list, e.g., Burst general.
- *Interconnection network*: Depending on the characteristics of the network connecting the `Edge nodes` to the `cloud`, it is possible to model it with a dedicated component, e.g., a *delay* station, with the mean service time and variance collected directly from the network.
- *Allocation/Deallocation of servers*: A policy similar to that described for the dynamic allocation of the servers to the `Edge nodes` can be used for their *deallocation*. In this case, a new *guard value* of the `Response time` of the `Edge nodes`, i.e., its *minimum* mean value, must be defined by the users and set in the autoscaler component. When reached, a server of the node can be deallocated and its load redistributed among the remaining ones or simply stop to load it.
- *Fluctuations*: Depending on the environments considered, the traffic of arriving scans can be affected by *fluctuations* with very high peaks and deep valleys. In these cases, to avoid problems of *instability* of the number of servers of the `Edge nodes`, it can be useful to define for each of the two guard values used by the allocation/deallocation policies of the autoscaler, ranges of tolerated values instead of the two mean values only.

6.2   Autoscaling Load Fluctuations

tags: open/closed, two classes, Source/Queue/Place/Transition/Sink, JSIMg.

In this case study we describe a multi-formalism model [5, 20] (with Queueing Networks and Petri Nets stations integrated) that simulates an autoscaler component that manages congestion created by fluctuations in incoming traffic and computational demands. The focus is on the description of the *dynamic routing mechanism* (that is state-dependent) of the arriving requests as a function of the load fluctuations of an online web service center. The solution described allows *cost savings*, in terms of resources used, while preserving the expected *system performance* and can be applied with considerable savings to exascale data centers.

6.2.1   Problem Description

Most data centers of Internet Service Providers experience load fluctuations caused by the combined effects of variability in incoming traffic rate and the computation time of the requests [13]. Depending on the service provided, fluctuations can have very different intensities and time scales. For example, in e-commerce sites, the increase of load due to seasonal sales can last several weeks with medium intensity and quarterly frequency, while unexpected events, such as special offers, create high spikes in requests with short duration and heavy computation time.

We can basically distinguish between *long-term* and *short-term fluctuations.* The former have low frequency, small/medium intensity and are generated by the typical growth trend of workloads. The latter have a short duration, high intensity and can occur at unpredictable times.

In such a variable scenario, the *right-sizing* problem, i.e., the identification of the minimum number of resources that must be used to achieve the performance objectives, is a very difficult problem. *Over-provisioning* may result in a waste of resources and money. On the other hand, *under-provisioning* can lead to violating customer expectations in terms of Quality of Service (QoS) with negative effects on business. *Autoscaling techniques* are increasingly used to dynamically allocate and release resources both in clouds (e.g., AWS Auto Scaler [1], Microsoft Azure autoscaler [27]), and in private data centers (e.g., [18, 22, 30]). Basically, these dynamic scaling techniques (usually divided into *horizontal* and *vertical* scaling techniques) monitor one (or more) *performance indicator* and when its *target value* is reached (or approached) trigger decisions to adapt the number of resources as the load increases or decreases. In the following we consider only the increase case because is the most critical for performance and furthermore the decisions taken in the decrease case are usually the opposite of those made in the first case.

When the target value of the performance indicator is detected, *horizontal scaling* typically allocate new resources while *vertical scaling* increases the capacity share of the resources.

Horizontal scalers provide good results when used with loads subject to long-term fluctuations such as those generated by physiological workload trends, whose growth rate increases progressively and continuously. But their application to loads subject to short-term fluctuations is unsatisfactory. The presence of load spikes has a very negative impact on performance as it creates a sudden congestion of resources which is responsible for the high Response times. Furthermore, they can foster horizontal scalers to make contradictory decisions in a short time that could generate dangerous *oscillations* in the number of resources provided. These unstable conditions must be avoided as much as possible as resources allocation are costly and time-consuming operations.

To address these drawbacks, we designed a *hierarchical scaler* (see, e.g., [33, 34]) with two operational layers shown in Fig. 6.8. The objective of the horizontal scaler at *Layer 1* is the typycal one: to provide the *minimum* number of resources (referred to as Web Servers) that should be used to achieve the performance target. This scaler has been enhanced with a second operational layer, *Layer 2*, consisting of a Spike Server that allocates CPU capacity to execute *load spikes* according to a *vertical scaling technique*. A request can be executed by a Web Server or by the Spike Server depending of the load conditions.

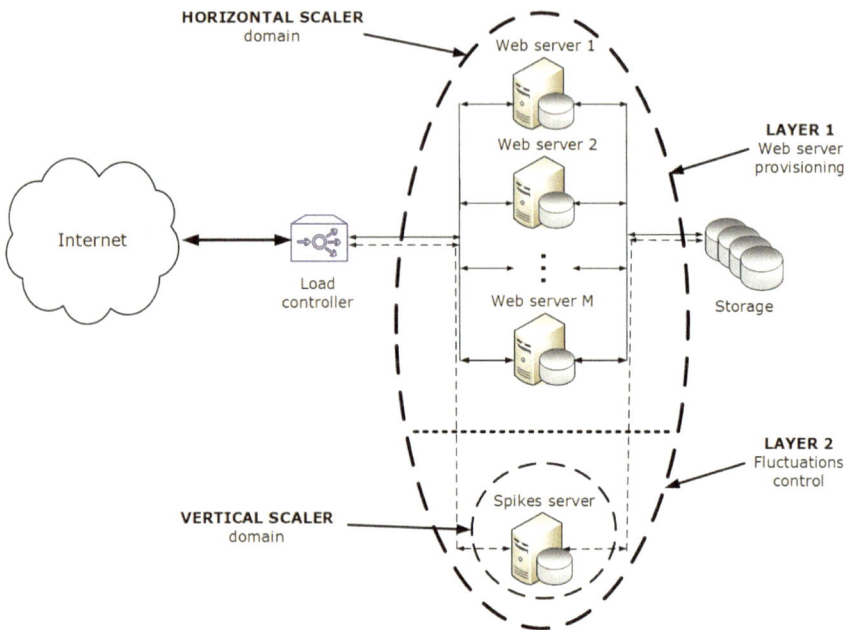

Fig. 6.8 Hierarchical autoscaler for load spikes

At *Layer 1*, a new `Web Server` is allocated when the monitored performance indicator reaches, or is close to, its *threshold* value. To make decisions on whether to scale or not, we have considered the performance indicator *mean* `Response time` R_0 of the data center, i.e., the mean time required by the execution of a request. *Layer 2* operations are triggered when a *load spike* is *expected* to arrive at one of the `Web servers`.

While the evaluation of R_0 is a well defined process, the load spikes are usually not so easy to predict with reasonable accuracy. Instead of running complex and time-consuming analyses on the traces of arriving requests, we considered the signals that anticipate the arrival of a potential peak load. More precisely, we consider a *Spike Indicator (SI)* metric whose alarm threshold SI^{max}, when reached, indicates that a peak load is likely to occur. Since a spike is anticipated by an increase in load in the system, we associate *SI* with the *number of concurrent requests in execution* in the considered `Web Server`. This metric is very suitable for the autoscaler as it is easy to measure and can detect the creation of peaks in their early stages not only in the arrival traffic flow but also in the request execution times. When *SI* reaches, or approaches, the threshold SI^{max}, a scaling decision should be made quickly to alleviate the congestion of the `Web Server`: the new incoming requests are routed to the `Spike Server`. As a consequence, the load of `Web Server` will decrease as running requests complete their execution. The routing of the requests is switched back to `Web Server` when *SI* decreases below SI^{max}. To minimize the fluctuations in the number of resources allocated, a range of values can be considered that includes SI^{max} instead of a single value (that we consider for simplicity).

Clearly, the detection of the correct value of SI^{max} is a very critical operation for the effectiveness of the autoscaler. If too many false positive spikes are detected the Spike Server tends to be congested. On the other side, if too many false negative spikes are detected the mechanism fails to reduce the congestion of Web Servers. The SI^{max} value is influenced by the characteristics of the workload, both by the arrival patterns and by the execution times, and by the performance objectives. In the following we will describe one of the possible approaches to tune SI^{max}.

The presence of the Spike Server has a *very positive impact* in reducing the System Response Time R_0. In fact, the larger values of Response times, mainly due to the congestion states of the Web Server, are replaced with smaller values obtained from the executions of the Spike Server, which is typically not congested. This *smoothing effect* reduces the variance of Response times, their mean value, and therefore the number of scale actions and their oscillations. The efficacy of the introduction of Spike Server is related to the following basic principle that applies to *open* models: the increase in Response time due to an increase $\Delta\lambda$ of load is greater than its decrease due to the same decrease $\Delta\lambda$. This effect is due to the vertical asymptote to which the Response time tends as the queue component approaches saturation.

The operating steps of the hierarchical autoscaler are:

1. at *Layer 1*, the horizontal scaler monitor the performance metric System Response Time R_0 and triggers congestion management actions when a threshold value has been exceeded. The value of R_0 is computed applying a moving window technique whose duration is a function of the characteristics of the workload. In the computation of R_0 both the execution times of the Web server and those of the Spike Server must be considered. According to the rules set at design phase, when the alarm threshold of R_0 is reached, or approached, the scaling decisions concerning the provisioning of new Web servers must be activated.

2. the control of the arrival of load spikes is *always active*, *Layer 2*, through the monitoring of the number of requests SI concurrently in execution in the Web Server. When the alarm threshold SI^{max} is reached, the dynamic routing to the Spike Server of new arriving requests is activated. When SI falls below SI^{max}, the incoming requests will be routed again to the Web Server. To avoid fluctuations, a *range* of tolerated values can be adopted instead of a single value SI^{max}.

3. if the System Response Time R_0 does not drop below its alarm threshold with the spikes control, then it is necessary to activate the actions triggered by the rules set in the autoscaler (typically increase in the number of servers). A further decrease of R_0 can be obtained by *vertical scaling* actions applied to the Spike Server increasing the share of the CPU dedicated to the application (when this is possible).

In this case study we focus on the model of the workload fluctuations and on the identification of the alarm threshold SI^{max} for the control of load peaks. Among the problems that can be studied are:

– evaluation of the influence of fluctuations in arriving requests with different time scales and intensities on the system performance
– impact of variability of service demands of requests on performance metrics
– assess the influence on System Response Time R_0 of the alarm threshold value SI^{max} for significant changes in workload arrival rate, e.g., up to about 40,000 req/h per Web Server
– identification of the value SI^{max} that minimizes the System Response Time R_0 for a given workload
– behavior of the autoscaler (in terms of the number of scaling up actions) with respect to the size of the moving window considered for the computation of the metrics used as performance indicators (e.g., the System Response Time, the Utilization of the Web Server and of the Spike Server)
– effects of vertical scaling actions of the CPU share of Spike Server on the number of servers provisioned as a function of arrival rates.

6.2.2   Model Implementation

The implemented *multi-formalism* model consisting of both *Queuing Networks* and *Petri Nets* components is shown in Fig. 6.9. Since this case study is focused on the autoscaling of load fluctuations, below we will concentrate on the description of *Layer 2* operations. At *Layer 1* the *horizontal scaler* performs the typical provisioning actions of new servers when the performance indicators exceed the threshold values (see, e.g., Sec. 6.1) balancing the load between them according to the policy adopted.

To simplify the presentation, we have introduced some assumptions that have small or no influence on the validity of the results. First, we modeled the app with only its *most utilized resource*, i.e., the bottleneck, that has a deep impact on the performance. The error introduced on the performance indexes ignoring the other resources should be very low as they are usually utilized much less than the bottleneck. Indeed, in many real-world cases, several important tasks of an app are allocated on a single (or very few) host server, typically very powerful and the most secure, which quickly becomes congested as the workload increases (e.g., the tasks that execute the front-end modules, the catalog and the cart services, the management of encryption/decryption keys for the payments, the 3D-secure procedure for online shops). The resource of the model that executes the requests is the *queue station* referred to as Web Server1. This is the resource that is replicated by the *Layer 1* autoscaler provisioning actions. The Spike Server at *Layer 2* is dedicated to the execution of the spikes of load.

Furthermore, to better investigate the behavior of the spike control, we have considered in the model *only one* server (Web Server1) with the connected Spike

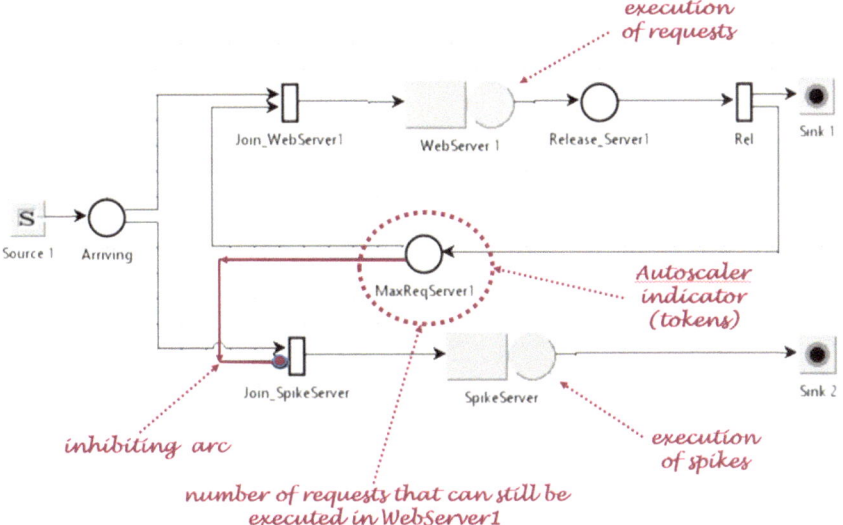

Fig. 6.9 Model with one `Web Server1` and one `Spike Server` for the auto-control of fluctuations

`Server`. Clearly, the results obtained for this initial configuration, with a web server and a spike server, also apply to each web server in the data center (if there are more than one), regardless of their number. Indeed, all servers can be considered independent of each other as their arrival rates are computed by the horizontal scaler algorithm, which is typically executed by a dedicated resource. Since the CPU capacity of a `Spike Server` is shared among several `Web Servers`, it is necessary to apply adequate scaling up actions on them as the number of `Web Servers` increases.

The workload consists of *two classes* of customers: the *incoming requests* submitted by the users of the application, and the *tokens*. The arriving *users requests*, referred to as `ArrivReq` and represented with an *open class*, are generated by the `Source1` station and routed to *place* `Arriving`. The *tokens*, referred to as `maxReqLink1`, are modeled with a *closed class* and are associated to the requests in execution (the SI metric), one token per request. Their maximum value represents the maximum number of requests that can be executed concurrently by the `Web Server1` (referred to as *alarm threshold* SI^{max}) for the load spikes control. At the beginning of the simulation, all tokens are located in *place* `MaxReqServer1`. The transition `JoinWebserver1`, see Fig. 6.10a, is enabled when a request arrives in *place* `Arriving` and there is at least *one token* available in *place* `MaxReqServer1`. At each activation, a request is sent to `Web Server1` and the number of available tokens in *place* `MaxReqServer1` is decremented by one. When a request is completely executed, *transition* `Rel` routes it to `Sink1`, see Fig. 6.11b, and returns the token to the *place* `MaxReqServer1` increasing the number of requests that can be in execution by one. When the number of tokens in `MaxReqServer1` is *zero*, the maximum number of requests in execution SI^{max} is reached and the autoscaler

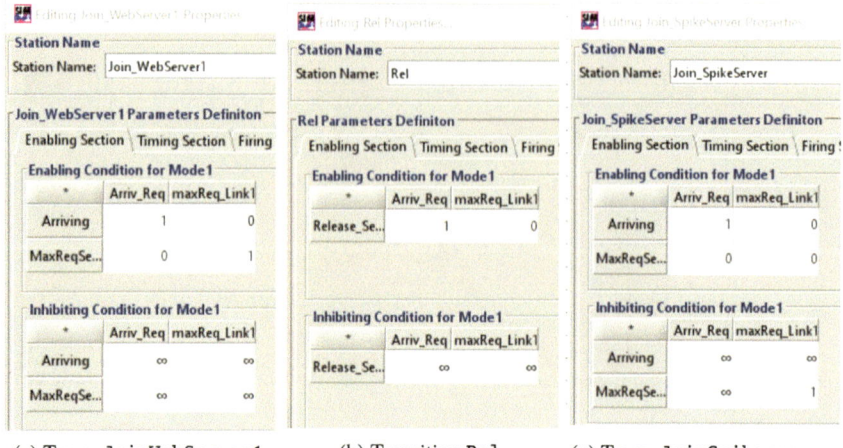

(a) Trans. `JoinWebServer1`   (b) Transition `Rel`   (c) Trans. `JoinSpikeserver`

Fig. 6.10 `Enabling` and `Inhibiting` conditions of the three `transitions`

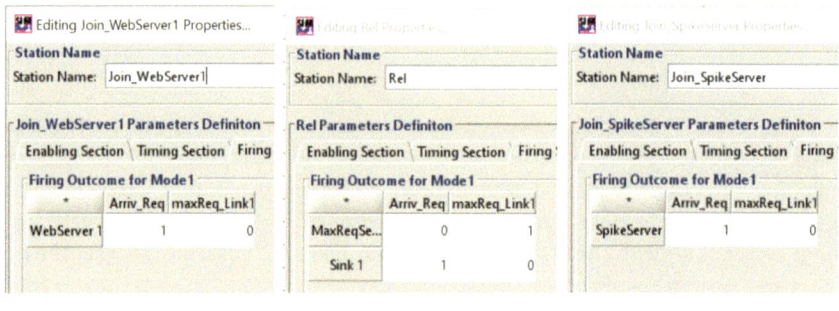

(a) Trans. `Join WebServer1`   (b) Transition `Rel`   (c) Trans. `JoinSpikeServer`

Fig. 6.11 `Firing` rules of the three `transitions`

control *routes* new arriving requests to the `Spike Server`, i.e., the *transition* `JoinWebServer1` is no longer activated. This is achieved through the *inhibiting* arc from *place* `MaxReqServer1` and *transition* `JoinSpikesServer`, see Fig. 6.10c. The value 1 in the inhibiting conditions of this *transition* means that when there are *one* or more tokens in *place* `MaxReqServer1` the transition is *blocked*. The values ∞ that appear in the inhibiting conditions indicate that the correspondent inhibitions are never met. To allow the computation of several interesting metrics, e.g., the `Response times` and the Throughput of the spikes, the requests executed by the `Spike Server` are addressed to the dedicated *sink* `Sink2`. The firing rules, i.e., the Throughput, of the three *transition* stations are shown in Fig. 6.11.

To reproduce the *fluctuations* of the incoming traffic, the distribution of the inter-arrival times of requests has been assumed *hyperexponential* with coefficient of variation cv = 4 and mean 0.15 s. The high variability of the service demands of

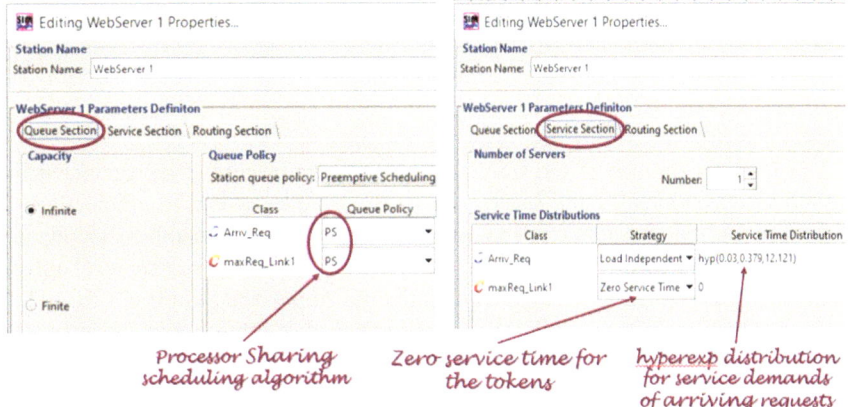

Fig. 6.12 Parameters of the Web Server1 station

the requests was modeled in both servers with a *hyperexponential* distribution with coefficient of variation cv = 4 and mean 0.16 s (see Fig. 6.12). The service demands of the tokens are set to zero to not interfere with the execution of arriving requests. The scheduling discipline of the two queue stations modeling the CPUs with two classes of customers is Processor Sharing (PS). This discipline is commonly used for the simulation of the time quantum policy of processors that share the capacity among all the requests to be executed, which can belong to different classes of workload.

The objectives of the case study required the execution of different types of analysis. To analyze the behavior of the model, several single simulation runs were performed with the collection of traces (see, e.g., Figs. 2.10, 2.11) with the CSV values of the performance metrics over time. The usual capacity planning problems are solved with What-if analyses using various control parameters, e.g., the *arrival rate of requests* generated by Source1, and the value of the alarm threshold SI^{max} of spike control.

6.2.3   Results

Of all the possible objectives that can be achieved with the implemented model, we will describe the operations required by the following four:

— **Obj.1**: *Implementation of the model of the autoscaler with two operational layers and evaluation of the correctness of its dynamic behavior to control load spikes.*

— **Obj.2**: *Given the arrival rate of requests of 400 req/min, evaluate the impact of different alarm thresholds SI^{max} of* Web Server1 *on performance metrics.*

— **Obj.3**: *Evaluate the behavior of* `System Response Time` R_0 *as the work-load grows to approximately 40,000 req/h.*
— **Obj.4**: *Analyze the impact of vertical scaling of* `Spike Server` *capacity on* `System Response Time` R_0.

Obj.1 shows the use of the CSV traces of performance metrics generated by the model executions for the analysis of its dynamic behavior. To efficiently use autoscaling techniques, it is very important to know the impact that the performance indicators monitored by autoscalers have on the satisfaction of service level agreements (SLAs). For example, what is the influence of the *Spike Indicator SI* on the `System Response Time` R_0? *Obj.2* and *Obj.3* address this issue. The impact of vertical scaling actions of CPU share of `Spike server` on the number of scaling actions is described by *Obj.4*

The description of the operations required to achieve the four objectives follows.

— **Obj.1: Model implementation of an autoscaler component that detects load peaks in** `Web Server1` **and relieves its congestion by dynamically routing new requests to a** `Spike Server`.

The structure of the model is described in the previous section.

To analyze the dynamic behavior of the model and to assess its correctness we collected the CSV traces with the values of several metrics over time (see Figs. 1.8, 2.10, 2.11). Several simulations were carried out using controlled workloads of increasing complexity. A visual evidence of the correctness of the load controller is provided in Fig. 6.13 that plots the `Number of requests` in execution in `Web Server 1` and in `Spike Server` over time. An interval of time of 1080 s, from 120 to 1200 s, has been considered. The alarm threshold SI^{max} (i.e., the max number of requests in execution in `Web Server1`) is set to 140 req. This value corresponds to the number of tokens of the closed class `maxReqLink1` that at the beginning of the simulation are in the *place* `MaxReqServ1`. Figure 6.13b shows that when SI^{max} is reached, e.g., in the interval 320–490 s, the state-dependent control of the autoscaler routes the new incoming requests to `Spike Server`. As soon as some requests complete their execution in `Web Server1`, the *SI* indicator drops below 140, e.g., in the interval from 570 to 690 s, and then the new incoming requests will be directed to `Web Server1` again.

The impact of load fluctuations on the performance are clearly shown in Fig. 6.14 which plot the `Response times` of `Web Server1` and `SpikServer` for the interval 120÷1200 s. For example, towards the end of a long period of high-load, at about 480 s, a peak of 50 s of the `Response time` of `Web Server1` occurs, see Fig. 6.14a. As expected, the `Response times` of `Spike Server`, see Fig. 6.14b, are much lower than those of `Web Server1`. A significant decrease of mean `System Response Time` R_0 with the workload considered, can be obtained simply by decreasing the alarm threshold SI^{max}. This action tends to balance the utilizations of the two servers, decreasing the congestion of `Web Server1` while increasing the load of `Spike Server` (see the following objectives).

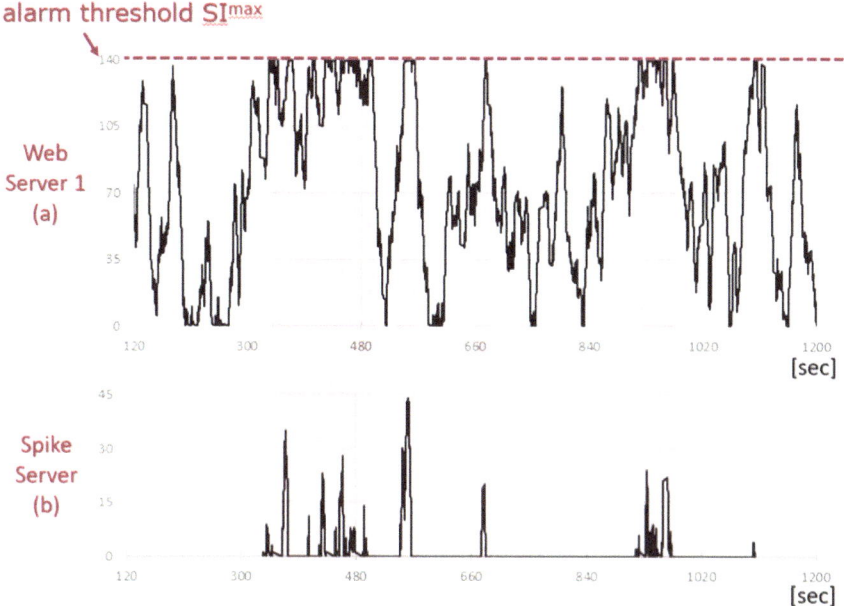

Fig. 6.13 Number of requests in Web Server1 (**a**) and Spike Server (**b**) in the interval $120 \div 1200$ s with the alarm threshold SI^{max} of *Spike Indicator* set to 140 req (initial population of place MaxReqServer1) and Arrival rate of 6.66 req/s

— Obj.2: With the arrival rate of 400 req/min (6.66 req/s), compute the performance indexes of Web Server1 **and** Spike Server **and the** System Response Time R_0 **for the alarm thresholds** SI^{max} **ranging from 10 to 160 req. Identify the value of** SI^{max} **that should be provided as input to the autoscaler in order to obtain a mean** System Response Time **as close as possible to the target value of 8 s.**

The *parameterization* of the workload is shown in Fig. 6.15. The flow of arriving requests (open class Arriv_Req) is generated by the Source station with a *hyper-exponential* distribution of Interarrival times with mean 0.15 s corresponding to the arriving rate of 6.66 req/s and coefficient of variation cv = 4. The service times of the two servers are *hyper-exponentially* distributed with mean 0.16 s and cv = 4 to simulate the fluctuations of service demands. The global population of the closed class maxReq_Link1 corresponds to the value of the alarm threshold SI^{max} for Web Server1 (in Fig. 6.15 it is $SI^{max} = 100$ req). The value of SI^{max} represents the maximum number of requests that can be in execution on Web Server1 which, once reached, identifies a state of *high-load* which causes the routing of arriving requests towards the Spike Server.

To tune the autoscaler parameters we evaluate the effects of the alarm threshold SI^{max} on the System Response Time. We used a What-if analysis with control parameter SI^{max} ranging from 10 to 160 req with increments of 10, so

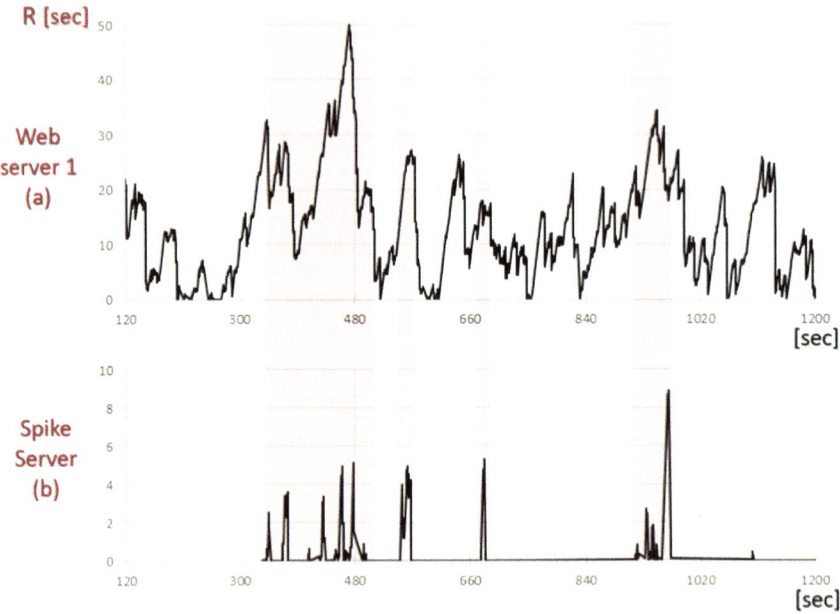

Fig. 6.14 `Response times of Web Server1` (**a**) and `Spike Server` (**b**) in the interval 120÷1200 s . with alarm threshold SI^{max} of *Spike Indicator* set to 140 req and arrival rate of requests 6.66 req/s

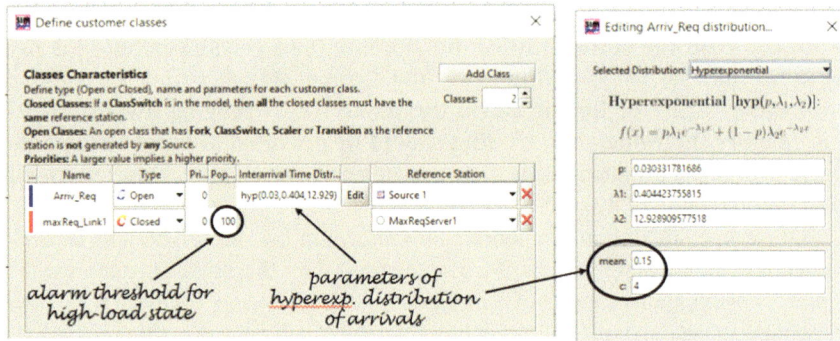

Fig. 6.15 Parameters of the `Source` station for the generation of the arriving flow of requests of 6.66 req/s and coefficient of variation cv = 4 (open class `Arriv_Req`), and setting of the alarm threshold $SI^{max} = 100$ req

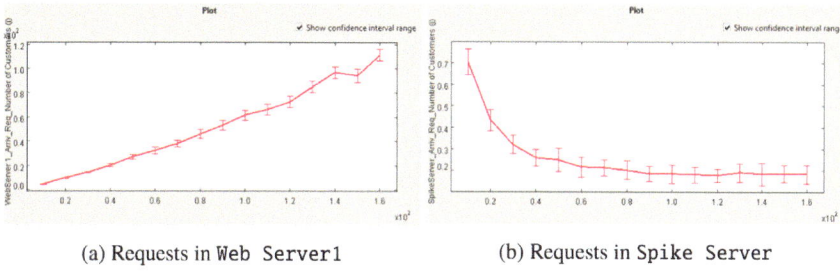

(a) Requests in `Web Server1`   (b) Requests in `Spike Server`

Fig. 6.16 Requests in execution *versus* alarm thresholds SI^{max} with load of 6.66 req/s

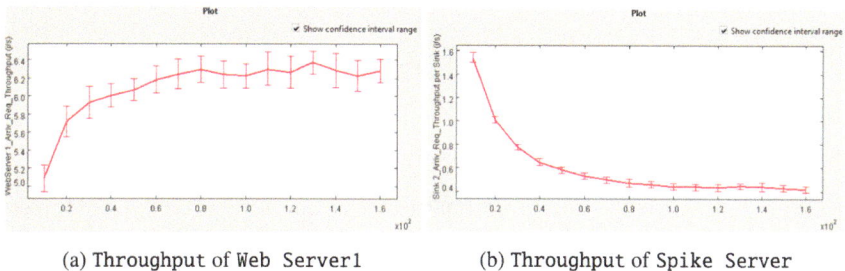

(a) Throughput of `Web Server1`   (b) Throughput of `Spike Server`

Fig. 6.17 Throughput of `Web Server1` and `Spike Server` *versus* SI^{max} (from 10 to 160 req)

overall 16 models are executed in sequence. We have considered such a wide range of values in order to provide a large set of data for the *training set* of the *machine learning* algorithm that will be applied in a second phase of the project. Some of the indexes detected are reported in Figs. 6.16, 6.17, 6.18. The number of times SI^{max} is reached decreases as its value grows from 10 to 160.

As SI^{max} increases, the Number of requests in execution, the Throughput and the Response time of Web Server1 increase while the corresponding indexes of Spike Server decrease. Let us remark that with the arrival rate of 6.66 req/s the Utilization of the Spike Server decreases from 0.4 (with $SI^{max} = 10$) to 0.12 (with $SI^{max} = 160$). This low Utilization is the motivation of the modest decrease of its Response times (see Fig. 6.18b) with the increase of SI^{max} from 10 to 160 req. The Utilization of Web Server1 increases almost linearly from 0.66 to 0.94 as SI^{max} increases. This is due to the scaling algorithm that route the arriving requests of Web Server1 dynamically to the Spike Server when SI^{max} is reached. It is important to note that we are evaluating the behavior of these performance indexes by keeping the request arrival rate *fixed* (6.66 req/s and, as seen above, not particularly high), so the saturation effects are very limited. In the following *Obj.3*, we will evaluate the system performance with different Arrival rates and the effects of server saturation on the System Response Time will be discussed.

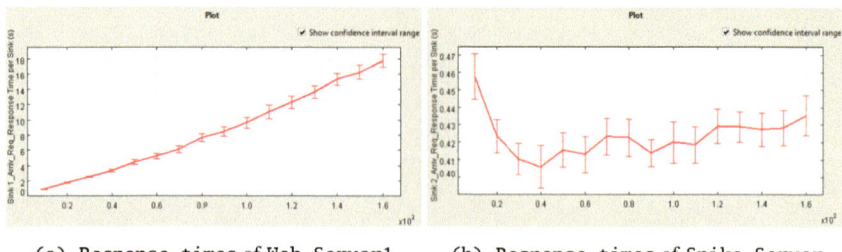

(a) **Response times** of **Web Server1**   (b) **Response times** of **Spike Server**

Fig. 6.18 Response times of Web Server1 and Spike Server *versus* SI^{max} (from 10 to 160 req)

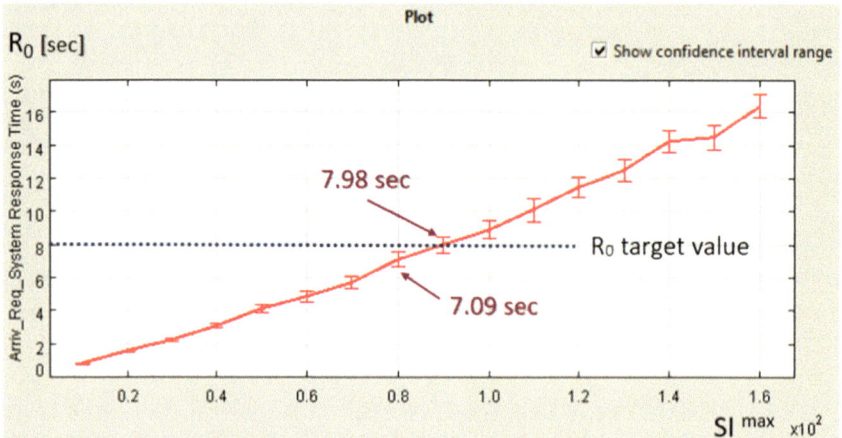

Fig. 6.19 System Response Time vs Alarm threshold SI^{max} with arrival rate of 6.66 req/s, the Interarrival times are *hyper-exponentially* distributed and cv = 4

The mean System Response Time R_0 of the model in Fig. 6.9 is given by the sum of the mean Response times of Web Server1 and Spike Server weighted by the respective percentages of System Throughput X_0:

$$R_0 = R_{WebServer1} \frac{X_{WebServer1}}{X_0} + R_{SpikeServer} \frac{X_{SpikeServer}}{X_0}$$

.

To identify the value of the alarm threshold SI^{max} that with the Arrival rate of 400 req/min (6.66 req/s) provide a System Response Time $R_0 \leq 8$ s a What-if analysis that performs repeated executions with SI^{max} as a *control parameter* ranging from 10 to 160 (globally 16 models) has been utilized.

As shown in Fig. 6.19, with $SI^{max} = 90$ the mean System Response Time R_0 is 7.98 s, too close to the target value of 8 s. A conservative answer to the question of *Obj.2* is $SI^{max} = 80$ that provides $R_0 = 7.09$ s.

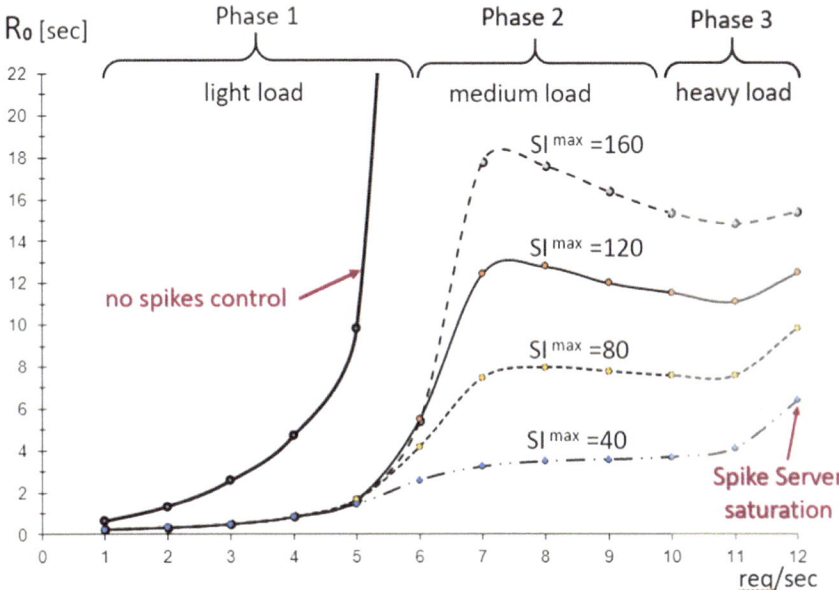

Fig. 6.20 `System Response time` *vs* `Arrival rate` (`Interarr.time` with *hyper-exp* distr. and cv = 4)

— **Obj.3: Assess the impact of various alarm thresholds** SI^{max} **on** `System Response Time` R_0 **for significant changes in workload arrival rate, from 1 to 12** req/s (**43,200** req/h).

To achieve this goal, a `What-if` analysis was performed for SI^{max} values from 40 to 160 req with arrival rates as *control parameter* ranging from 1 to 12 req/s (60 to 720 req/min). The positive impact of the dynamic control of the high-load states of `Web Server1` on `System Response Time` R_0 is highlighted in Fig. 6.20. The lower curve represents R_0 for $SI^{max} = 40$ req, while the upper one for $SI^{max} = 160$ req. These two values correspond respectively to the minimum and the maximum utilization of `Web Server1`. In Fig. 6.20 three different *operational phases* can be identified according to the workload intensity: *light, medium, heavy*.

In *Phase 1*, which includes arrival rates between 1 and 6 req/s, R_0 is less than 6 s for all the SI^{max} values. It is important to note that *without any scaling action* the R_0 corresponding to an arrival rate of 5 req/s is about 10 s, while small increments of successive arrivals cause its *very large* increases. This is because the `Web Server1` with these arrival rates and *no spike control* is highly utilized (the `Response time` grows to infinity) and we are approaching the `Throughput` bound. Let us remind that the maximum arrival rate that `Web Server1` can process when it is saturated, is given by $\lambda_{max} = 1/D_{WebServer1} = 6.25$ req/s (from the utilization law $U_{WebServer1} = \lambda_0 \, D_{WebServer1}$).

The *low* arrival rates of *Phase 1* reduced drastically the need of autoscaling actions, and thus the load of `Spike Server`. As a consequence, the contribution of the Spike Server to the System Response Time is minimal. In fact, even towards the extreme of the interval with the highest arrival rate of 6 req/s and $SI^{max} = 160$ req, the utilization of Spike Server is very low and thus its executions have minimum impact on the computation of $R_0 = 5.42$ s. The correspondent utilization of Web Server1 is $U_{WebServer1} = 0.93$.

Phase 2 includes arrival rates between 6 and 10 req/s and shows increasing values of R_0 until the arrival rate of about 8 req/s. This increase is mainly due to the increment of the arrival rate at the Web Server1 that now is close to congestion. A further increase in arrival rate from 8 to 10 req/s generates an increase in the number of high-load states of Web Server1 detected by the autoscaler and therefore the number of requests routed to Spike Server grows progressively. As a consequence, R_0 is decreasing since the contribution to its computation due to the Spike Server executions, which are much shorter than those of Web Server1, becomes more substantial as its Throughput increases (with a medium utilization). Indeed, for example, with $SI^{max} = 160$ req the $U_{SpikeServer}$ is 28% with 8 req/s arrival rate and is 60% with 10 req/s. The correspondent Throughputs $X_{SpikeServer}$ are 1.76 req/s and 3.77 req/s, and the Response times $R_{SpikeServer}$ are 0.69 s and 1.54 s, respectively. R_0 will return to growth as the utilization of Spike Server increases and therefore its Response times increase (in the *Phase 3*).

It should also be emphasized that the pattern of the requests arriving to the Spike Server is typically bursty since in most cases consists of load spikes, see e.g. Fig. 6.13b, and it is known that the presence of bursts has a very negative influence on performance.

Phase 3 is characterized by two factors: the *heavy* workload (between 10 and 12 req/s) and the *congestion* of the Spike Server. Considering, for example, $SI^{max} = 160$ req, the utilization of the Spike Server $U_{SpikeServer}$ for arrival rates of 10 and 12 req/s are 60% and 97%, respectively, and its Response times $R_{SpikeServer}$ are 1.54 and 7.54 s. Since the corresponding Throughputs $X_{SpikeServer}$ are 3.77 and 5.69 req/s (representing approximately 50% of the System Throughput X_0), the impact on the mean System Response Time R_0 ?? of Spike Server executions becomes substantial. As the SI^{max} values decrease from 160 to 40 req the increases of R_0 become more evident as the load of Spike Server increases.

The values shown in the previous figures are very important for setting the parameters of the autoscaler, and for a *machine learning* algorithm, in order to satisfy the target value of the selected performance metric. For example, consider an arrival rate of 9 req/s with cv = 4 and a target value of the scaling metric $R_0 \leq 8$ s. From Fig. 6.20 it is possible to see that with the alarm threshold of 80 req this objective can be achieved. Note that the autoscaling policy tries to use a web server as much as possible as long as the specified target value of the scaling metric is met. With an increase of the arrival rate from 9 to 12 req/s this target value cannot be matched. In this case, a scaling action is needed. If the Spike Server has unused capacity, a vertical scaling can be activated (see the following *Obj.4*) increasing the share devoted to the application. If this is not possible, then a new server must be allocated (through a scaling action at *Layer 1*) to handle the service demands.

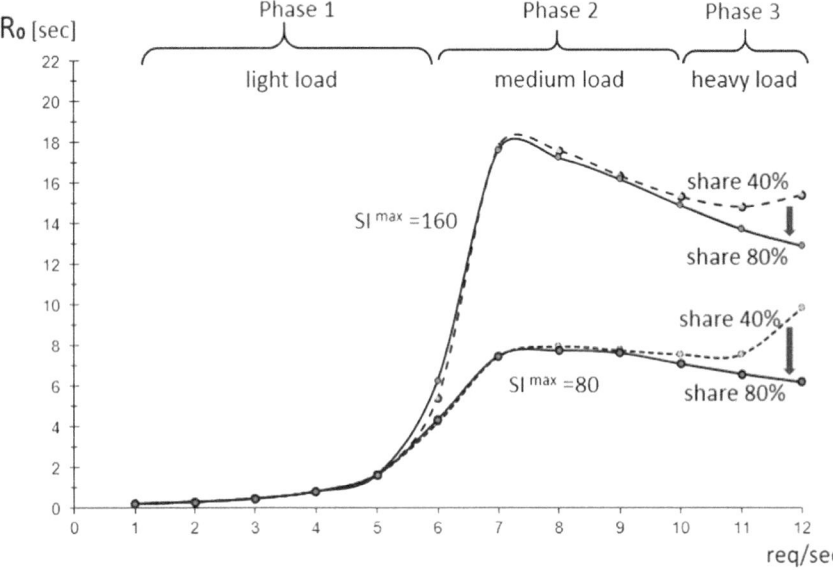

Fig. 6.21 Impact of doubling the CPU capacity share of `Spike Server` (vertical scaling from 40% to 80%) on `System Response time` R_0 with 80 and 160 req alarm thresholds

— Obj.4: Assess the impact on `System Response Time` of a vertical scaling action that double the capacity share of `Spike Server` from 40% to 80%.

When the `Spike Server` approaches congestion state (as in *Phase 3* of Fig. 6.20), it causes a degradation of *System Response Time* R_0. In this case, before activating the horizontal scaling actions by increasing the number of web servers, it can be very effective to apply a *vertical scaling* action by increasing, *if possible*, the *CPU share* of the `Spike Server` dedicated to the application. This vertical scaling action is typically much less expensive than the horizontal one and faster to apply. Clearly, in this case the CPU power of the `Spike Server` must be greater (at least two times or more) than that of the `Web Server1`.

Since in the previous *Obj.3* the CPU share was 40%, in this *Obj.4* we evaluate the effects on R_0 obtained by doubling this share to 80%. The parameter of the model that must be changed is the *service demand* $D_{SpikeServer}$ of the `Spike Server` that must be set to 80 ms instead of 160 ms. The R_0 values for arrival rates between 1 and 12 req/s and the alarm thresholds SI^{max} 80 and 160 *req* are shown in Fig. 6.21.

The dashed lines represent the R_0 values with the original CPU share of 40% (considered in the previous Objectives) while the solid lines represent the corresponding values with the CPU share doubled to 80%. As expected, significant decreases in R_0 values are achieved in the *Phase 3* area where the `Spike Server` is more utilized. For example, with arrival rate of 12 req/s and $SI^{max} = 80$ req the target value $R_0 \leq 8$ s considered in *Obj.3* can be reached with 80% share ($R_0 = 6.2$ s), while with 40% share this is *not possible* ($R_0 = 9.83$ s).

6.2.4   Limitations and Improvements

- *Various application scenarios*: The case study described is focused on the implementation of a model that exhibit dynamic behavior as a function of the load characteristics. With simple modifications/upgrades, it can be used in various application scenarios, such as, for example, to model the *dynamic load split* between the servers of a *private* and a *public* cloud, or to evaluate the performance *impact* of the *number of cores* in the various partitions of an HPC system.
- *Oscillations control*: To *minimize* the oscillations in the number of provisioned resources, it would be better to use a range of values as a target for the scaling indicators rather than just the mean values.
- *Workload with heterogeneous apps*: The modeling approach described can also be used with *multiclass workloads*. The rules for enabling, firing, and inhibiting can be specified for each individual class.
- *Vertical scaling*: For a given arrival rate of requests, the effects of *vertical scaling* of `Spike Server` can be investigated with a `What-if` analysis that uses as *control parameter* its service demands scaled according to the CPU share policy adopted.
- *Horizontal scaling*: The model described can be enhanced with the implementation of the *horizontal scaling provisioning policy* at *Layer 1*. The structure used for `Web Server1` must be replicated for each of the considered `Web Servers` whose load is controlled by a new `transition` component that implement the replication policy.
- *Machine Learning*: Efficient scaling policies in complex scenarios can be obtained integrating several techniques, like *modeling, and machine learning* into a *single tool* that dynamically tune the parameters according to the varying load conditions. For example, from the results of a sequence of models obtained with a *What-if* analysis, a *machine learning* algorithm can derive the set of parameters that keep performance indicators as close as possible to their target values.
- *Use of Finite Capacity Region*: The model described can also be implemented using a `Finite Capacity Region` (with max capacity set to SI^{max}) for each `Web Server` and implementing the *firing rule* in the *transition* that manages the flow of arriving requests according to the planned scheduling policy.

6.3   Simulation of the Workflow of a Web App

tags: open, three classes, Source/Queue/Class-Switch, JSIMg.

While in the typical Queueing Network models the paths followed by the requests between the stations are defined according to probabilistic rules, with the use of the `Class-Switch` parameter of the requests it is possible to describe in JSIMg the paths in a *deterministic* way. A similar behavior can be modeled also using the Petri Nets stations (see, e.g., Sect. 6.2).

6.3.1 Problem Description

Regardless of the paradigm adopted in modern web application architectures (e.g., web services, microservices, serverless) software developers must describe the business logic of the apps through workflows representing the sequence of execution of the tasks. Depending on its layout, mapping a workflow to a queuing network model may not be an easy task. More precisely, we refer to the case in which a request after being executed by a station and flowed through the model, returns to that station and requires service times and routing very different from the ones required previously. The problem we face arises because JMT does not store the execution history of a request in terms of paths followed between the various resources. To solve this, we use the *class identifier parameter Class ID* associated with each running request to track only its recent execution history.

In fact, each request in execution is assigned a *Class ID* that is used to describe its behavior and characteristics, such as, type (open or closed), priority, mean and distribution of service times. Routing algorithms are defined on a per-class basis. Of fundamental importance to the problem approached is that a request may change *Class ID* during its the execution flowing through a Class-Switch station or when a specific routing algorithm is selected. Therefore, with the use of the *Class ID* parameter we can know the last station visited by a request and the path followed.

To describe this technique we consider a simplified version of the e-commerce application of an online food shopping company. The web services of the software platform are allocated on two powerful servers, referred to as Server A and Server B, of the private cloud infrastructure. Figure 6.22 shows the layout of the data center with the paths followed by the requests and the relative Classes. The sequence of execution of the paths for each request coincides with the numbers of the *Class IDs*.

Server A is a multicore system that execute several services of Front-End. Among them are: customer authentication, administrative and CRM processes, interaction with the payment service (for the strong authentication for payments), checkout operations with the update of the DB, invoice generation, shipping and tracking services, and update of customer data. Server B is a multiprocessors blade system, highly scalable, fault tolerant, with redundant configuration for continuous availability, equipped with large RAM memory and SSDs storage for the DBs. Among the most important services allocated are those for browsing the catalog, processing the shopping cart, and managing the DBs of products and customers. To provide the minimum Response time to customers, an in-memory DB is implemented to dynamically cache each customer's most recent purchases.

A third server Server P, located in the data center of an external provider, is used for *payment services.*

To reduce the complexity of the description we have considered a simplified version of the workflow of the e-commerce app, see Fig. 6.23, consisting only of the services that are needed to describe the problem approached and its solution. Figure 6.23 shows the services considered and the servers where are stored.

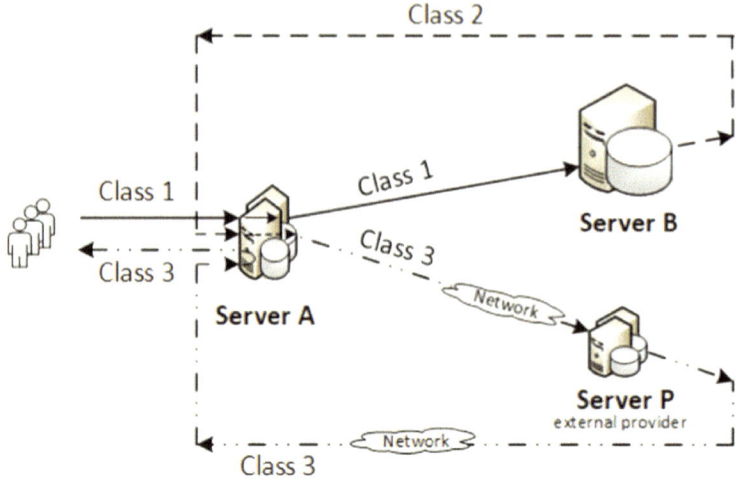

Fig. 6.22 The data center with the path followed by the requests and their *Class IDs*

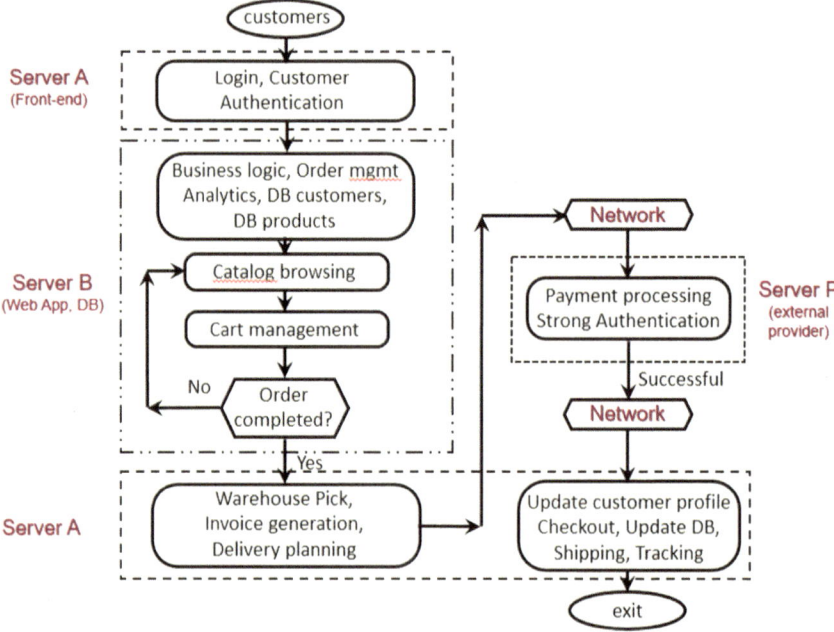

Fig. 6.23 Short version of the workflow of an order submission to an online grocery store

According to the business logic of the e-commerce app, the complete execution of a request requires *three visits* to Server A, one to Server B and one to Server P. At each visit to Server A, different web services are executed which require different mean service times. The sequence of visits to the three servers A, B, and P during a complete execution is A-B-A-P-A. With a smart use of the *Class IDs* we may model this deterministic routing of the requests among the servers.

In addition to the implementation of the model for executing the workflow of Fig. 6.23, the capacity planning study requires:

– the performance forecast of the e-commerce app with the *current* workload and the one-factor authentication level for payments for a wide range of arriving requests;
– the impact assessment of a new web service for the *Strong Customer Authentication* (SCA)
  Indeed, according to PSD2 (Payment Services Directive EU) a new authentication service is planned to replace the current one to enhance the security of online payments with a *two-factor authentication* levels;
– the Throughput bound of the current system and the actions to be applied to process a workload 15% higher than the current one (with max arrival rate of about 5000 req/h).

6.3.2   Model Implementation

The model implemented with JSMg is shown in Fig. 6.24. We use the *Class ID* of the requests to trace the path between stations followed during their executions. The sequence of paths modeled is shown in Figs. 6.22 and 6.23. The arriving requests from Source station, generated with Class1 as *Class IDs*, are sent to Server A. After the execution of the services scheduled for this *first visit* to Server A, the requests are routed to Server B and then to Class-Switch station CS. This station, which has zero service time, change the *Class IDs* of incoming requests to new ones according to the probabilities described in its parameters. In our case, see Fig. 6.25, the *Class IDs* of the requests arriving from Server B (that are Class1) are changed to Class2 before being redirected to Server A.

After the execution of the services scheduled for this *second visit* to Server A (which are different from those executed in the first visit that were for Class1 requests) the Class2 requests are routed to the payment server Server P of an external provider. At the end of this service, the *Class IDs* of requests are changed to Class3 by Class-Switch CS station that route them back to Server A. The service demands of this *third visit* to Server A are those for the requests of Class3 type. Then, the routing algorithm sends them to the Sink station where they exit the model.

In the model implemented we have not *explicitly* represented the network connections to payment server Server P and the User think times. Indeed, the service times of the network components, typically modeled with Delay

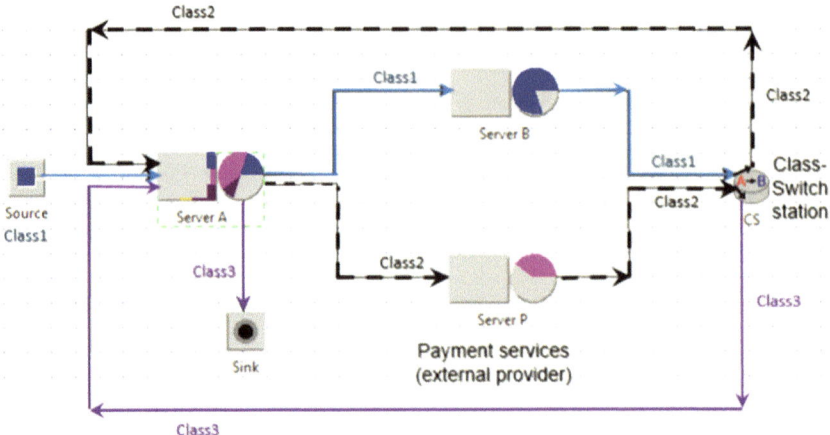

Fig. 6.24 The JSIMg model implemented

Fig. 6.25 Class-Switch
probabilities of the CS
station

CS Parameters Definiton			
Class Switch Matrix \ Routing Section			
CS Strategies			
*	Class1	Class2	Class3
Class1	0.0 (0%)	1.0 (100%)	0.0 (0%)
Class2	0.0 (0%)	0.0 (0%)	1.0 (100%)
Class3	0.0 (0%)	0.0 (0%)	1.0 (100%)

stations, are negligible compared to the service times of the other components of the model and therefore their impact on the performance is practically zero. As for the User Think Times it should be emphasized that with this type of e-commerce app, related to online grocery shopping, their values, especially those on the *browser-side* (i.e., between the selection of products), are highly variable from user to user as they are deeply influenced by the characteristics of individual customers (e.g., age, type of network connection, digital equipment used). Thus, having a reliable forecast of their values and distribution is practically impossible and somewhat useless. Furthermore, not considering the User think times increases the reliability of the metric System Response Times R to evaluate the differences between the various versions of web services and security protocols. We must keep in mind that in this case we simulate the *worst case* scenario related to the stress of the resources since the load is the *maximum* possible.

The Service times required by all services executed during each visit have been parametrized with their global Service demands D*s*. *Three* different workloads should be considered: the *current*, (called *light workload*), with an average of

Table 6.2 `Service demands` [s] to the servers of the current workload with one-factor (left) and with two-factor (right) authentication for payment security

Stations	Classes			Stations	Classes		
	1	2	3		1	2	3
Server A (Login, Front end, ...)	0.2	0.4	0.1	Server A (Login, Front end, ...)	0.2	0.4	0.15
Server B (Web App Serv., DBs, ...)	0.8	0	0	Server B (Web App Serv., DBs, ...)	0.8	0	0
Server P (Payment Provider)	0	0.4	0	Server P (Payment Provider)	0	0.7	0

25 products in the shopping cart per customer in each session and one-factor authentication for secure payment, the same workload with a new two-factor authentication system, and the new expected workload (called *heavy workload*) with a 15% increase in incoming traffic compared to current one. Tables 6.2 shows the `Service demands` of the first two.

The fluctuations in the number of items purchased per session and service times are considered in the distributions of the global service demands Ds. The values in the boxes are those modified by the two-factor payment system. To model the fluctuations in the number of items and in service times required by their different types we have assumed the exponential distribution of service demands Ds. If necessary, it is possible to select distributions with the same mean and greater variance, for example hyper-exponential, with a single click in the station parameterization windows. The scheduling discipline of the servers is PS, processor sharing. The traffic intensities analyzed range from 0.5 to 1.2 req/s (about 4300 req/h).

6.3.3 Results

In what follows we will describe the activities regarding the following three objectives:

— **Obj.1**: *Implementation of a model to execute the workflow of Fig.6.23*

— **Obj.2**: *Capacity planning of the data center with the current workload and evaluation of the impact of a two factor authentication system*

— **Obj.3**: *Computation of the* `Throughput` *bound and prediction of the performance of a new heavy workload that has a max arrival rate of 5000 req/h*

— **Obj.1: Implementation of a model to execute the workflow of the e-commerce app with deterministic paths**.

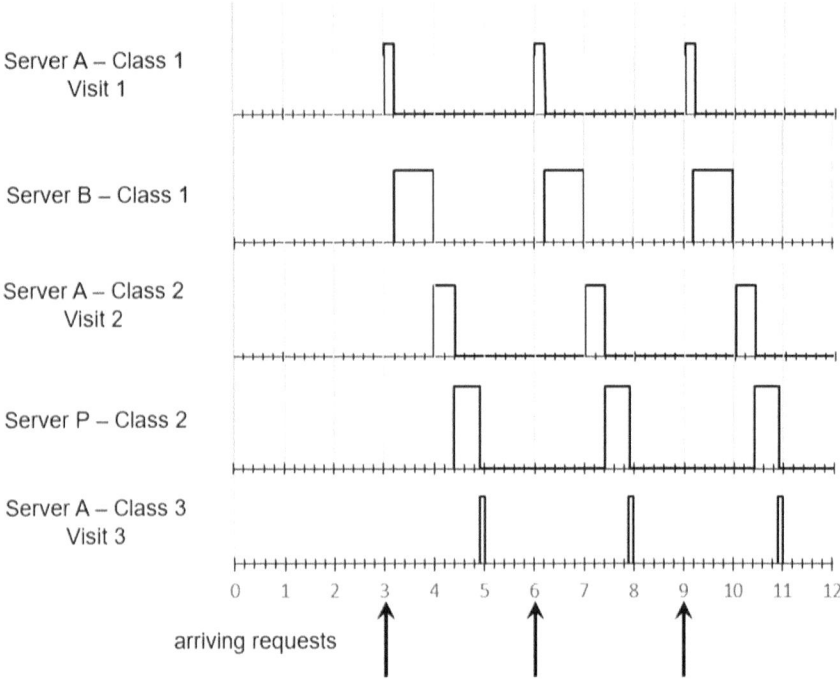

Fig. 6.26 Temporal sequence of visits to the three servers during the execution of a request

The workflow with the tasks that are executed for an online order submission is shown in Fig. 6.23. To construct a simple example that allows the visual evidence of the sequence of visits to the three servers A,B, and P we have implemented the model of Fig. 6.24 assuming that all the parameters have constant values. In this example: the interarrival times are 3 s, the service demand of the *first* visit to Server A is 0.2 s (the request is of *Class1*), the service demand to Server B is 0.8 s (the request is of *Class1*), the service demand of the *second* visit to Server A is 0.4 s (the request is now of *Class2*), the service demand to Server P is 0.4 s (the request is of *Class2*), and the service demand of the *third* visit to Server A is 0.1 s (the request is now of *Class3*).

The temporal diagram of Fig. 6.26 provides visual representation of the *sequence of visits* A-B-A-P-A to the three servers during the complete execution of a request. The values plotted in this diagram are obtained simply by flagging the check-box Statistical Results (Stat.Res.) of the correspondent *performance index* selected in the Performance Indices window (see Fig. 1.8) (the CSV files with the values of the *Response Times* will be generated automatically).

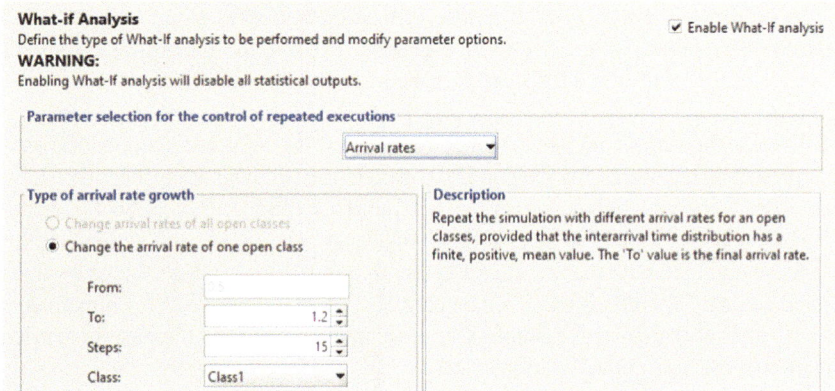

Fig. 6.27 Execution of 15 models with `Arrival rates` from 0.5 to 1.2 req/s

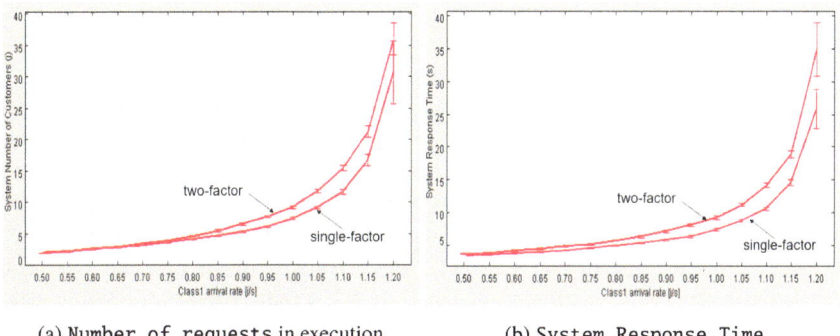

(a) `Number of requests` in execution   (b) `System Response Time`

Fig. 6.28 `Number of requests` N in execution (**a**) and `System Response` times R [s] (**b**) with one-factor and two-factor authentication layers of security

— Obj.2: Capacity planning of the data center with the current workload and evaluation of the impact of a two-factor authentication system for secure payments.

The traffic intensities considered range from 0.5 to 1.2 req/s (4320 req/h). The `What-if` analysis of Fig. 6.27 execute 15 independent models with arrival rates increasing by 0.05 each model . Figure 6.28 show the `Number of Requests` N in concurrent execution and the `System Response times` R with the `Service demands` of the current workload with *one-factor authentication* security system (see Table 6.2).

The values of N range from 1.4 to 30.81 req while those of R range from 2.9 to 25.8 s. Recall that, as already pointed out, the values of R do not include `User Think times` both at the *browser* and the *session* levels.

The *bottleneck* is `server` B whose utilization increases from 0.39, with 0.5 req/s, to 0.95, with 1.2 req/s. Its *service demand* is 0.8 s, the maximum among all servers, and therefore the `Throughput` bound of the system is $X_0 = 1/D_{max} = 1.25$ req/s.

To detect the impact on performance of the new *two-factor* authentication system for secure payments we executed the `What-if` analysis with the `service demands` of the new payment service (see Table 6.2). Figure 6.28 allow the visual comparison of the values of N and R obtained with the two-factor authentication with those obtained with the one-factor authentication. With 1.2 req/s the new values of N and R are 35.9 req and 34.8 s, respectively. The 9 s increase in R compared to the old value obtained with a single-factor is mainly due to the increase in service demands of the new authentication system.

— Obj.3: Computation of the `Throughput` bound and performance prediction of a new heavy workload that has a max arrival rate of about 5000 req/h.

The current workload intensity is expected to increase by approximately 15% following the acquisition of a new online grocery store company. Arrival rates with a maximum value of about 5000 req/h are also expected.

The bottleneck with the current workload is `Server` B which constrains `Throughput` to be at most $1/D_{max} = 1.25$ req/s, which is less than the new required maximum 1.4 req/s. Among the possible actions to improve the `Throughput` bound it has been decided to replace the current `Server` B with a new one that is twice as fast (equipped with new processors, more cores, and larger RAM). As a consequence, the service demand of `Server` B is 0.4 s, half of the previous one. The new D_{max} is the one of `Server` A, equal to 0.7 s, that becomes the bottleneck. So, the new `Throughput` bound is 1.42 req/s, which satisfies the constraint of 5000 req/h. Figure 6.29 shows the `System Response time` R of the two data center configurations with the *old* (upper curve) and *new* (lower curve) `Server` B respectively.

It must be pointed out that the performance gains obtained with the new `Server` B that is twice as fast of the old one are not as expected, e.g., the `Throughput` bound increased of about 14% only. Indeed, with the new `Server` B the limit to the extent of the improvement is imposed by `Server` A which has become the *new bottleneck* with the second highest service demand of the original data center, i.e., 0.7 s, as it was the *secondary bottleneck*.

6.3.4 Limitations and Improvements

- *Workload characterization*: identifying *multiple classes* of customers (rather than a single one as done in the case study) may be better for providing customers with more accurate `Response times` with respect their characteristics (in terms of the number of products purchased).

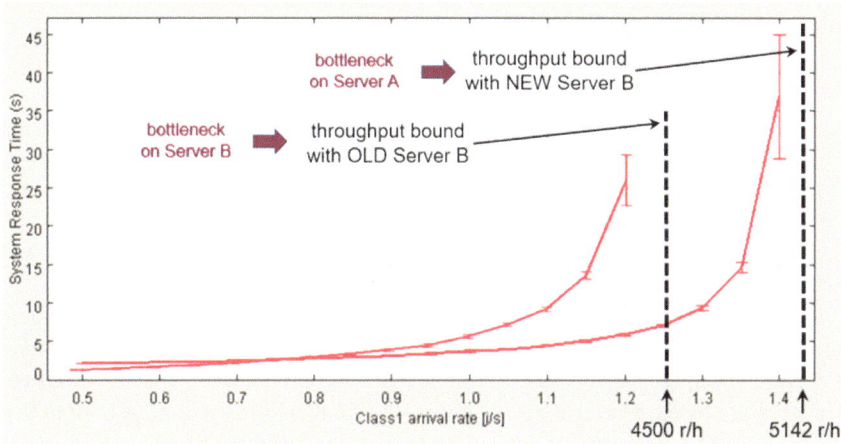

Fig. 6.29 `System Response time R` [s] of the two data center configurations

- *Flow of incoming customers*: the pattern of arriving requests can be simulated with high precision capturing the fluctuations with some of the distributions implemented in JMT, e.g., `Hyper-exponential, Coxian, Phase-Type, Burst, Markovian Arrival Processes`) or using the `Replayer` to replicate the data collected from a real workload.
- *Load balancing*: regardless of the paradigm adopted in web application architectures, the identification of elementary tasks, their dependencies on other tasks and their allocation among web services, are the actions that play a *key role* in the load balancing of servers in a data center.

6.4 A Crowd Computing Platform

tags: open/closed, multiple class, Source/Delay/Queue/FCR/Sink, Exp/Hyper-exp, JSIMg.

This case study describes an application of a simple but powerful structure that can be implemented with one of the JSIMg features: the *Finite Capacity Region* (FCR). It can be used either stand alone, as described below, or as a part of more complex models [29], for example to simulate the servers downtime (due to failure or other causes of shutdown) in large data centers, to control the load to a set of servers, or to implement the zig-bee energy savings feature [7].

6.4.1 Problem Description

The *crowd paradigm* has been used for centuries to solve problems whose difficulty is beyond the capacity of single individuals or organizations: a *group* (i.e., the *crowd*) of subjects *cooperate* to solve a problem. With the evolution of digital technologies, and particularly the Internet, crowd applications encompass a wide range of real-world problems of both a scientific or non-scientific nature from agriculture, to health-care, funding, searching, social productivity, distributed weather forecast, problem-solving and ideas-sharing.

In this case study, we consider a crowd of individuals that collectively contribute with their digital devices (computers, servers, tablets, etc.) to the implementation of a large computing infrastructure. A device can be added or removed from the infrastructure by each contributor. The *members* of this infrastructure belong to two groups: *contributors* and *associates*. The former are authorized to add and remove their equipments to the infrastructure of the crowd that they can use free of charge. The latter can *only* use the infrastructure devices and are charged for their computations. Associate members have been introduced to increase the economic sustainability of the crowd, their number is larger than that of contributors. We will collectively refer to the members of the two categories as `users` and we assume that the service demands of both the categories are similar. In this ideal *crowd computing platform* (Fig. 6.30), the contributors receive by the crowd manager the app that allow them to add their computers to the platform, becoming a `node` accessible by the community, or to remove it. The crowd manager is responsible for the managing of the resources of the platform through dedicated servers. Among others, *scalability* is one of the important features that they exhibit. For their characteristics, these types of infrastructures can also be referred to as *open cloud computing systems*.

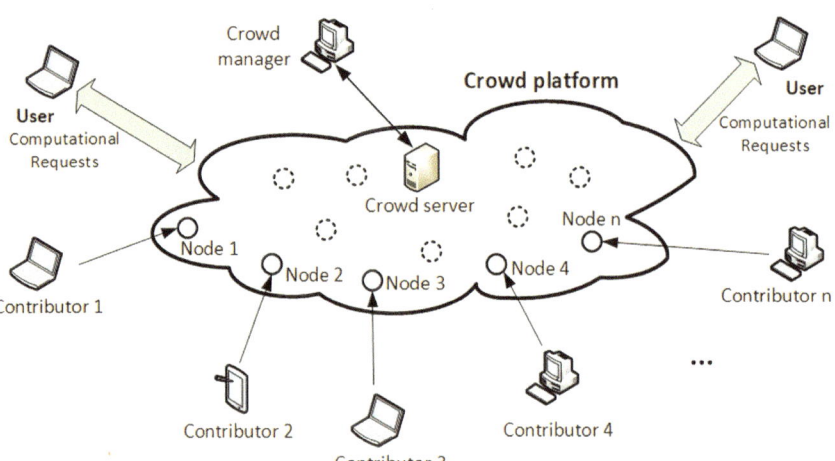

Fig. 6.30 Layout of the considered crowd computing scenario

In these applications, the processes of *contribution* (i.e., *arrivals*) and *removal* (i.e., *delete*) of the equipments to the platform are very peculiar and follow unpredictable distributions.

In the following sections, we focus on the simulation of these two processes and we analyze their impact on the performance of the crowd platform. More precisely, we evaluate the behavior of `System Response Times` of the user requests as a function of the variance of unaivalability time of the nodes.

6.4.2 Model Implementation

The model implemented with JSIMg is shown in Fig. 6.31. The flow of computational requests submitted by contributors and associates members, has been simulated with the requests of the *open* class `users` generated by the source station `Source1`. The `Service` demands of both the group of members have the same statistical characteristics, i.e., the same distribution (exponential) and the same mean $D_{user} = 4$ s. Thus we assume that all the computational requests belong to the same class.

The number of contributors is 200, each can add/remove a system that can execute the user requests. We simulate the computational devices of the platform, i.e., the *nodes*, with the 200 servers of the single queue station `CompServers`. To model the add/remove behavior of the 200 nodes, we use the `Finite Capacity Region` (FCR) `CrowdPlatform`, with capacity $N_{FCR} = 200$ customers, and a *closed* class `Node` with 200 customers. The `Node` customers flow through the *queue* station `Unavailable` and the *delay* station `Available`.

Thus, the workload of the model consists of *two* classes of customers: `User` (open) and `Node` (closed). In JSIMg the queue of requests entering an FCR is unique. In Fig. 6.31 two queues are drawn only for reasons of graphical representation.

Fig. 6.31 The crowd computing model: the `CrowdPlatform` region has a limited capacity, *class-2* `Node` customers have *higher* priority than *class-1* `User` customers

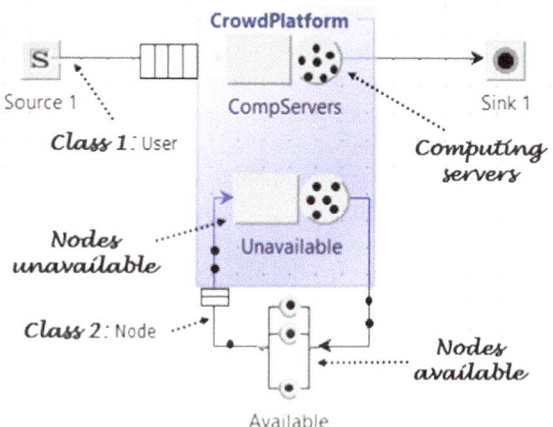

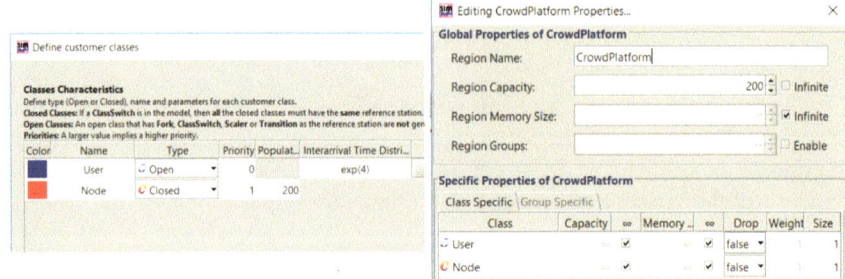

(a) Parameters of User and Node classes   (b) Parameters of FCR region

Fig. 6.32 User (open) and Node (closed with high priority) classes (**a**), and the FCR (**b**)

The *parameter settings* of the *two-class* workload is shown in Fig. 6.32a. The open class User describes the computational requests submitted by *all* the users, contributors and associates, arriving at the platform with rate $\lambda = 4$ req/s and *exponential* distribution of Interarrival times (whose mean is $1/\lambda = 0.25$ s). The closed class Node of 200 customers and priority 1 (higher than that of the User class) has been added to represent the systems of the platform that may be available/unavailable to execute the User requests.

Figure 6.32b shows the parameters of the FCR. The maximum Region capacity is $N_{FCR} = 200$ customers, including both user and node. The default values (infinite) of the maximum number of customers per class have no effects as in any case the maximum value of 200 customers in the FCR is a constraint that cannot be exceeded. The Drop policy is set to false for *both* the classes since we do not want to drop the requests (both User and Node) arriving when the FCR is *full* but we want to keep them in a queue waiting to be admitted inside.

The queue station CompServers, located inside the FCR, has a single queue and 200 servers, each server will execute a user request. Since the number of customers in the FCR is limited, i.e., it is $N_{FCR} = N_{FCR,User} + N_{FCR,Node} \leq 200$, any Node customer within the FCR (in the Unavailable station) decreases the number of servers available for user computations in the CompServers station.

The primary effect of *removing* a node is represented in the model by an increase of customers at the Unavailable station and thus a decrease in the number of servers available for computations at CompServers station. Similarly, the primary effect of *adding* a node is represented by a decrease of the Unavailable customers (a customer move to the Available station *outside* the FCR) and an increase in the number of servers available for computations at CompServers station. The result will be an increase in the node activity and an improvement in the platform performance.

The behavior of the model is as follows:

- if a `User` request arrives at `CrowdPlatform` when there is at least one server available in the `CompServers` station, i.e., when it is $N_{CompServers} + N_{Unavailable} < 200$, then it is executed immediately;
- if a `User` request arrives to `CrowdPlatform` when no computing server is available to users (i.e., when it is $N_{CompServers} + N_{Unavailable} = 200$) then it must wait until a user request complete its execution or a new node is added and that the eventual queue of requests already waiting for a server (i.e., in queue to enter the FCR) becomes empty;
- when a `Node` removal request arrives at `CrowdPlatform`, i.e., a `Node` customer is released by the `Available` station, and it is $N_{FCR} < 200$, then the number of servers available in the `CompServers` station is decreased by one unit;
- when a `Node` removal request arrives at `CrowdPlatform` and it is $N_{FCR} = 200$ then it must *wait* in queue to enter the FCR until a `User` request complete its execution and release the server or a new server is added (i.e., a `Node` customer exit the FCR). Indeed, in spite that the `Node` requests have higher priority than `User`, since the scheduling discipline of the queue of requests waiting to enter the FCR is FIFO *non-preemptive*, i.e., an arriving *removal* request of a node does not interrupt the execution of a `User` request but waits for its completion to lock the server. The requests in queue are served according to their `priority`.

In this case study we focus on the behavior of the number of nodes of the platform that are available/unavailable for computations. Typically, in this type of applications the time in which a node is *unavailable* follows an unpredictable distribution with a very large variance. This can be explained by considering that each contributor is independent of the others and follows custom working schedules. As a first approach, we consider the mean *unavailability* time $S_{Unavailable} = 1$ s (`Service time` of the `Unavailable` queue station, located inside the FCR), and *hyperexponential* distribution. Several models with the same mean `Service time` and different coefficients of variation cv are executed. Let us remind that to model a *hyper-exponential* distribution in JSIMg it is sufficient to set its mean value and coefficient of variation (see, e.g., Fig. 5.10). The *availability* times are modeled with the service times of the *delay* station `Available` (located *outside* the FCR), with mean $S_{Available} = 60$ s and *exponential* distribution.

6.4.3 Results

The simple model implemented allows to answer several capacity planning questions. For example: how does the platform `Response time` vary with the number of nodes? which is the impact on performance of the arrival rate of user requests and of the distribution of interarrival times? which is the bottleneck of the infrastructure which constraints the `Throughput`? what happens if we alleviate/remove the bottleneck? which will be the effect on `Response times` of an increase of the number of associates members? which are the scalability limits of the platform (hardware components capacity, software requests, distributions of service demands, variance of interarrival times, ...)?

In this case study we concentrate on the modeling of the addition/removal of nodes to the crowd platform. The primary effect of the interaction of these two processes is reflected by the dynamic changes in the number of nodes available/unavailable for the execution of user requests. As described in the previous section, we use the *exponential* distribution to model *availability times* and the *hyperexponential* distribution to model *unavailability times*.

Among the possible objectives of the capacity planning study, we describe in detail the following two.

Obj.1: Evaluate the impact of the variance of unavailability times of the platform nodes (keeping constant the mean value) on the `Response Time` of `User` requests.

To achieve the objective of the study we cannot use the `What-if` feature since in JSIMg the variance of a distribution is not one of the control parameters admitted. Thus, we ran five independent JSIMg models with different values of the variance of the unavailabity times. More precisely, for the `Service times` of `Unavailable` station we considered the same mean $S_{Unavailable} = 1$ s and five different coefficients of variation cv = *1, 5, 10, 15*, and *20* of their *hyper-exponential* distribution.

The graphs in Fig. 6.33 show the results of five models of Fig. 6.31 obtained with different cv of $S_{Unavailable}$. For each cv, the corresponding 99% confidence interval is also shown. As expected, the high variance of the *Unavailability* time causes a degradation in the performance. As the cv increases, the `System Response time` of `User` requests grows. For example, with cv = 20 the model yielded $R_{0,User} = 203.4$

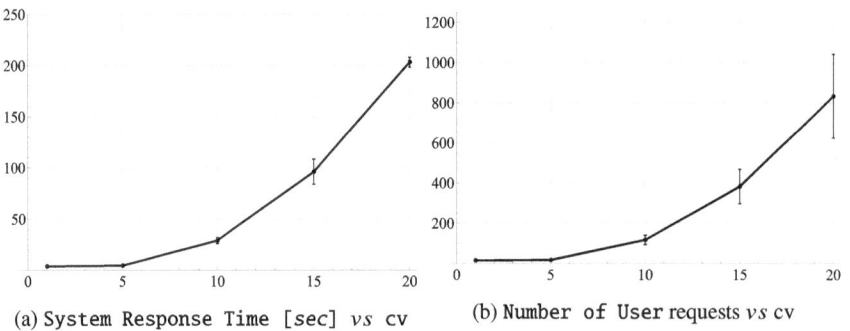

(a) `System Response Time [sec]` *vs* `cv`   (b) `Number of User` requests *vs* `cv`

Fig. 6.33 `System Response time` (**a**) and `System Number of User requests` in the platform (in execution and in queue for FCR) (**b**) *versus* coefficient of variation of *Unavailability* time

s (see Fig. 6.33a). Note that this index is at the `System` level because it also includes the *queue* time (if present) of the requests waiting to enter the FCR when all nodes are unavailable.

Similarly, Fig. 6.33b shows that with the increases of cv, also increases the mean number of `User` requests in the system. Note that this index is defined as `System Number of customers` because it includes both `User` requests that are submitted to the platform but that are queued waiting for a node (to enter the FCR) *and* requests that are in execution (inside the FCR). Indeed, its mean values can be higher that 200 (e.g., with cv = 20 it is $N_{0,User} = 834\ req$).

Obj.2: To answer some questions of the capacity planning study it is required a detailed statistical analysis of the values of three performance indexes: `Number of nodes` **available in the platform,** `Response time` **and** `Number of user` **requests arrived at the platform. The study of their behavior over time is also requested.**

To achieve this objective it is necessary to tick the checkboxes `Stat.Res.` in the `Performance Indices` definition window corresponding to the indexes analyzed. In Fig. 6.34 a statistical analysis is requested for the indexes `Number of customers` of `Available` station, `System Response times` of the `User` requests, and `System Number of Customers` because the corresponding three checkboxes `Stat.Res.` are checked.

For each selected index, a CSV file with all its values is generated. In Fig. 6.34b a sample of the CSV file generated by the `Number of customers` in the `Available` station is shown. The values of the three columns are: the *time stamps* of the event, the actual *number of customers* in the station, and the *time interval* since the last event, respectively.

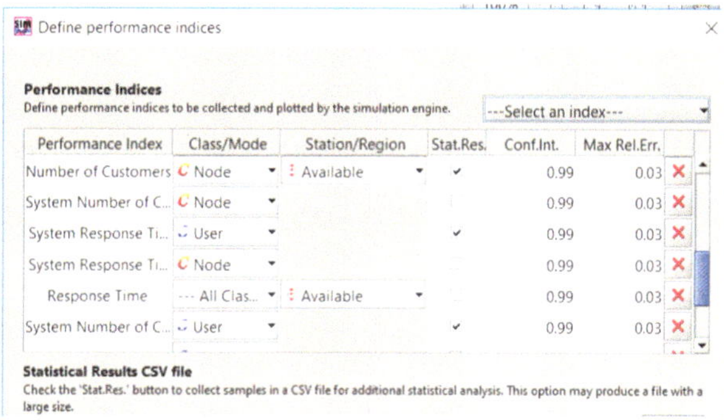

(a) Some indexes collected: three require a statistical analysis

965.6752	120	0.126594
965.7072	119	0.032
965.7134	118	0.006181
965.7316	117	0.018229
965.8153	118	0.083626
965.8741	117	0.058872
966.8751	116	1.000939
967.0792	115	0.204136
967.1746	114	0.095358
967.1878	113	0.013287
968.8305	112	1.642666
969.3437	113	0.513202
969.5126	114	0.16887
969.5143	113	0.001742
969.6332	114	0.118886
969.9327	115	0.299533

(b) `Number of Customers`

Fig. 6.34 Selection of statistical analyses (see `Stat.Res.` check boxes) of three performance indexes (**a**) and a sample of the CSV file of `Number of Customers` of the `Available` station (**b**)

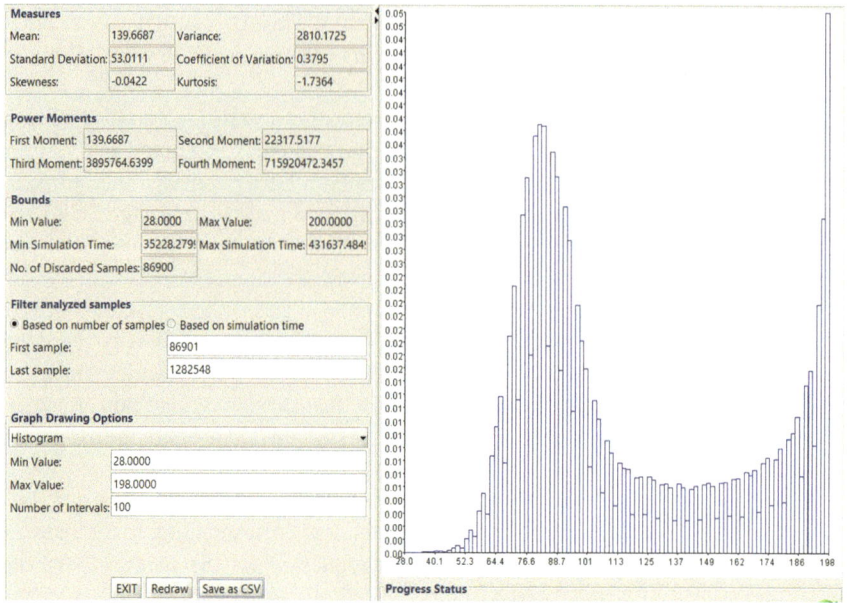

Fig. 6.35 Statistical indexes computed for the `Number of Unavailable` nodes

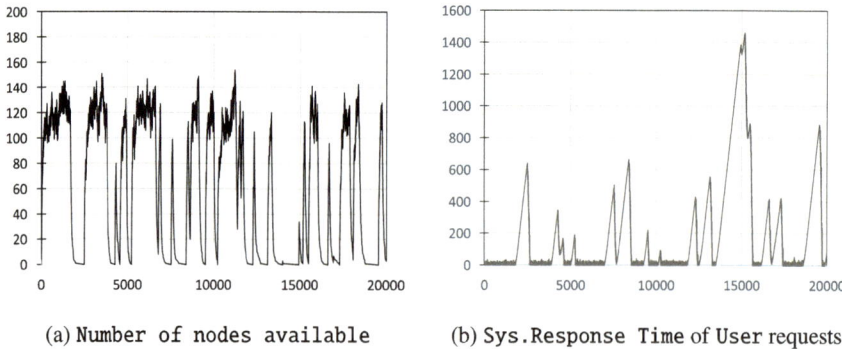

(a) `Number of nodes available`   (b) `Sys.Response Time` of User requests

Fig. 6.36 Behavior of the `number of nodes available` in the crowd platform (**a**) and of the `System Response times of User` requests (**b**) in the time interval $0 \div 20000$ s

For example, Fig. 6.35 shows the statistical indexes computed for the `Number of customers` in the `Unavailable` node. The `histogram` graph style has been selected.

By processing the CSV files there is the possibility to analyze the behavior of the indexes over time. Figure 6.36a represents the behavior of the `Number of Customers` of class-`Node` in the `Available` station (that correspond to the number of nodes available in the crowd platform) over time. Each increasing step means the occurrence of an arrival (a contribution) of a new system (`node`) or the completion of the execution of a `User` request that releases the server. Each decreasing step means that a system has been removed from the crowd or that a server has been assigned to a newly arrived `User` request. The values plotted have been obtained in a model with mean unavailability time $S_{Unava} = 1$ s and cv = 20.

Figure 6.36a shows for the same time interval the behavior of the `System Response time` of the `User` requests. The values of this index are highly fluctuating. As can be seen from Fig. 6.36, the high peaks of `Response times` (see, e.g., the one ending at about 15000 s in Fig. 6.36b) occur after periods where the number of nodes available for computations is very low or null (see Fig. 6.36a). In fact, the primary effect of these periods is represented by the fast increase of the queue of `User` requests waiting to enter the FCR, resulting in a significant increase in their `Response time`.

Appendix A
What May Be Useful to Know

A.1 Routing Probabilities Versus Visits to Resources

tags: open, single class, Source/Queue/Sink, JMVA/JSIM.

In many performance studies the *probabilities* that the requests follow given paths through the resources are known, while in other cases the number of *visits* that requests during their execution make to resources are known. In this section we derive the *relationships* between routing probabilities and visits.

Let us consider the open queueing network of Fig. A.1 that model a system with a central server, the Application server AS, and three Storage servers S_i in parallel. The probabilities p_i's that after a visit to the Application server AS a request is routed to Storage server S_i are known. The index 0 is used to represent the outside part of the queueing network, p_0 is the path that will be followed by a request that has completed its execution and leaves the model. To simplify the presentation, we assume that a request is routed to this path only once in his lifetime, so the number of visits V_0 that it performs outside the network is one. According to the structure of the network of Fig. A.1 it is $\sum_{i=0}^{3} p_i = 1$.

The requests arrive to the system with rate λ_0 and can be regarded as generated by a station external to the model representing the users. This station, used as Reference station, is visited only once during the execution of a request and it is used to compute the Response time R_0 and the Throughput X_0 of the *entire* system. Indeed, R_0 is defined as the period of time between the arriving to the system of a new request (leaving the Reference station) and its departure once completed (entering the Reference station), and X_0 is the rate of completed requests that enter the Reference station. Sometimes R_0 is also referred to as

© The Editor(s) (if applicable) and The Author(s) 2024
G. Serazzi, *Performance Engineering*,
https://doi.org/10.1007/978-3-031-36763-2

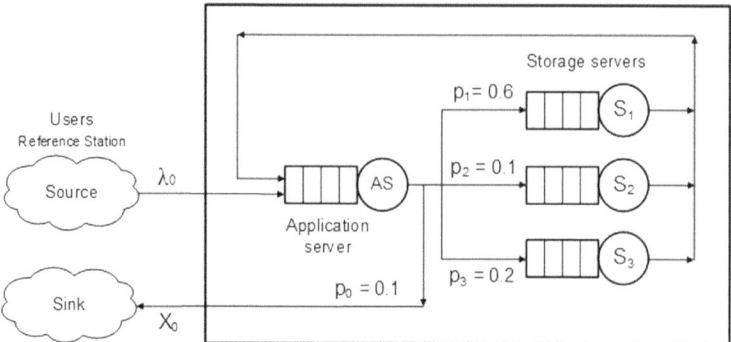

Fig. A.1 Open model with one `Application server` AS and three `Storage servers` S_i

Table A.1 Routing probability matrix for the model of Fig. A.1

	AS	$S1$	$S2$	$S3$
AS	0	p_1	p_2	p_3
$S1$	1	0	0	0
$S2$	1	0	0	0
$S3$	1	0	0	0

System cycle time since it represents the time required to a request to make complete cycle in the system with respect to the `Reference station`.

The routing probabilities of the queueing network of Fig. A.1 are shown in Table A.1. The element ij of this matrix represents the probability that after a visit to resource i a request is routed to resource j. Since a model in order to be solved analytically must be *flow balanced*, i.e., it must be in *equilibrium*, (see Sect.1.2) the number of requests that enter a station equals the number of requests that exit. We may say also that the throughput X_0 of the entire model, or that of a station, will be the same as the arrival rate λ_0 to the model, or to that station, respectively. Thus, for any station i of the queueing network it will be $\lambda_i = X_i$ (*flow in = flow out*). For the model of Fig. A.1 we may write the following system of equations:

$$
\begin{cases}
\lambda_0 + \lambda_{S1} + \lambda_{S2} + \lambda_{S3} = \lambda_{AS} \\
p_1 \lambda_{AS} = \lambda_{S1} \\
p_2 \lambda_{AS} = \lambda_{S2} \\
p_3 \lambda_{AS} = \lambda_{S3}
\end{cases}
\tag{A.1}
$$

Since it is $p_0 = 1 - (p_1 + p_2 + p_3)$, from Eq. A.1 we have

$$
\lambda_{AS} = \lambda_0 \frac{1}{p_0} \qquad \lambda_{S1} = \lambda_0 \frac{p_1}{p_0} \qquad \lambda_{S2} = \lambda_0 \frac{p_2}{p_0} \qquad \lambda_{S3} = \lambda_0 \frac{p_3}{p_0} \tag{A.2}
$$

From these results and recalling that the global number of operations C_i completed by station i divided by the observation interval T is its throughput X_i, we may derive the average number of visits V_{AS} that during its execution a request makes to station AS (recall that we *assume* $V_0 = 1$):

$$\frac{\lambda_{AS}}{\lambda_0} = \frac{C_{AS}/T}{C_0/T} = V_{AS} = \frac{1}{p_0} \tag{A.3}$$

where C_0 is the global number of requests executed by the system in T. Similarly, for the other stations it will be

$$V_{S1} = \frac{p_1}{p_0} \qquad V_{S2} = \frac{p_2}{p_0} \qquad V_{S3} = \frac{p_3}{p_0} \tag{A.4}$$

Dividing the first equation of (A.1) by λ_0 we obtain the following relation between a central resource and the peripheral resources it connects to in a network with a *central server*

$$V_{AS} = 1 + V_{S1} + V_{S2} + V_{S3}$$

that may be used in several models having complex structures.

Equations A.3 and A.4, with the values of the routing probabilities of Fig. A.1, provide the following visits to stations:

$$V_{AS} = \frac{1}{0.1} = 10 \qquad V_{S1} = \frac{0.6}{0.1} = 6 \qquad V_{S2} = \frac{0.1}{0.1} = 1 \qquad V_{S3} = \frac{0.2}{0.1} = 2$$

Let us note that, depending on the objectives of the performance study, as Reference station can be selected a station different from the one representing the Users, which is typically visited only once. In this case, the Response time and the Throughput of the system must be computed with respect to the new Reference station. Since typically its visits are greater than one, it is necessary to rescale the visits to all stations dividing their values by the original visits of the new Reference station. New values of visits must be computed (after the scaling, the visits to the new Reference station will become one), and the model provides the correct values of all the performance indexes computed with respect to the new Reference station.

A.2   Confidence Intervals

Random numbers are used in the simulation of computer systems to generate sequences of independent random values assigned to input parameters like Interarrival times, Service times, Think times, Service demands, Routing probabilities. Consequently, the output of simulations are sequences of random values describing metrics such as Response times, Number of customers, Throughput, Utilizations, and users have to estimate their distribution functions, their *mean values* and *variance*.

In Fig. A.2 a general view of a simulation process is shown. Several independent runs are executed, each run utilizes its own sequence of random values of input parameters and produce a corresponding sequence of output values of results.

A sequence of random numbers is constructed using a *seed* to initialize the generator algorithm. There is a direct correspondence between the values of seeds and the sequences generated. Since the selection of the seed is typically left to the generator algorithm, that usually derive its value from the clock time, it is clear that a different sequence of random numbers is generated at each execution. As a consequence, each execution of the same model will use a particular realization of the input sequences and will produce distinct realizations of the set of output sequences.

A user *must be able* to estimate which one of the values, for example a mean value, computed on the different output sequences of several independent simulation runs is the closest to the *true* mean value of the variable considered.

This is a classical statistical inference problem: using the results obtained from a *sample*, i.e., a subset of the elements of the analyzed population, a user need to derive conclusions at the level of the *entire* population.

For example, let us consider the variable describing the Response times R of a job computed by a model. The true mean value of R is unknown and we want to evaluate the accuracy of the mean computed on a sample of *n independent* executions of the same job. This type of estimation, that provides a single value (the mean of *n* values), is known as *point estimation*.

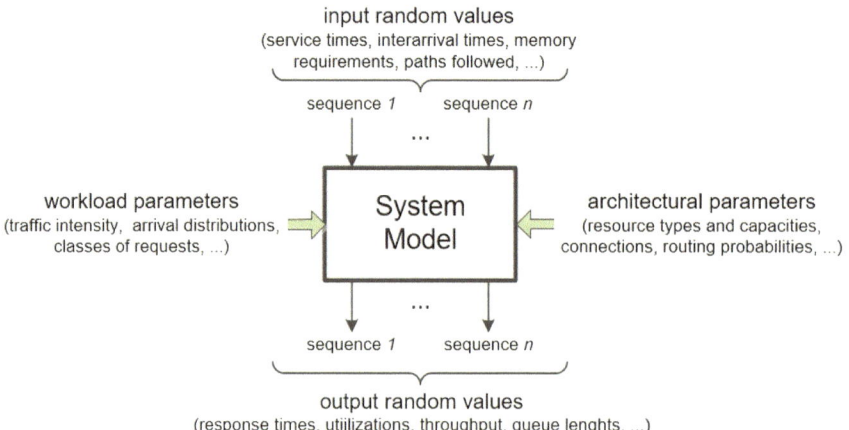

Fig. A.2 High-level view of a simulation process. Several runs are executed, each run utilizes its own sequence of input random values and generate the corresponding sequence of output results

Notice that *almost never* the point estimate coincides with the *true* value of the variable to be estimated. Indeed, it is not possible to know *how far* the value computed with a sample on n independent runs is from the true value of the index analyzed.

However, based on the values computed using the data collected by repeated *independent* executions, it is possible to construct an *interval estimate*, referred to as *confidence interval*, that will contain the true value of the index being estimated with a given probability $(1 - \alpha)$, referred to as *confidence level* $100(1 - \alpha)\%$, where α is the probability that the correct value of the parameter lies outside the confidence interval. Some of the more frequently used confidence levels are 90%, 95%, 99% (see, e.g., Fig. 1.8 for JSIMg).

It should be pointed out that when we consider **only one sample of size** n and we compute the $100(1 - \alpha)\%$ confidence interval, it represents the likely range for the correct mean value of the estimated parameter (that may or *may not* be contained). For example, if we consider the 95% confidence intervals generated from 100 samples (of the same size and collected independently) we may expect that 95 of them will contain (and 5 of them do not contain) the correct mean value.

We *assume* that the mean and the variance of the index to be estimated are *unknown*, and we consider *symmetric* confidence intervals (that generate intervals of *minimum* size). By the *Central Limit theorem* we may assume that the sample means computed considering *several* independent executions of size n (n must have a **large value** >30) of the same job *tend* to a normal distribution. Since the real value of the variance σ^2 is unknown, we may use its *estimated* value σ_s^2 computed with the values of the sample. With $n > 30$, a good approximation of the confidence interval $100(1 - \alpha)\%$ for the mean of the estimated index R is

$$\left(\overline{x} - z_{\alpha/2}\frac{\sigma_s}{\sqrt{n}}\right) < R < \left(\overline{x} + z_{\alpha/2}\frac{\sigma_s}{\sqrt{n}}\right) \tag{A.5}$$

where $\overline{x}$ and σ_s^2 are the mean and the variance computed with the sample values, $(1 - \alpha)$ is the confidence level (e.g., 90%, 95%, 99%) and $z_{\alpha/2}$ is a value that can be obtained from the table of the *standard normal distribution* function such that $P(Z < -z) = P(Z > z) = \alpha/2$. By definition, the area under the z density function from the value $-z_{\alpha/2}$ to its left is equal to $-\alpha/2$. Z is the standard normal $N(0,1)$ random variable $Z = (\overline{x} - mean)/(\sigma_s/\sqrt{n})$. For example, to compute the 95% or the 99% confidence intervals, the values of $z_{\alpha/2}$ are 1.96 and 2.576, respectively.

When the *number of executions* is **small** (the **sample size is n<30**), to have an acceptable estimate of the confidence intervals it is better to use the *Student's t-distribution* instead of the *standard normal* distribution. Thus, in Eq. A.5

instead of the $z_{\alpha/2}$ the $t_{n-1;\alpha/2}$ should be used. Indeed, the random variable $T = (\bar{x} - mean)/(\sigma_s/\sqrt{n})$ has a student *t-distribution* with *n-1* degrees of freedom. It is convenient to have sample sizes of at least 10. The values of $t_{n-1;\alpha/2}$ may be easily found in the tables of the Student t-distribution with $n - 1$ degrees of freedom.

A.3   Details on Reliability Models of Chap. 5

System with n-Parallel Components

Consider the reliability model of a system with n identical and statistically indepen-dent *parallel components*, with exponentially distributed lives, and non-repairable. This system fails when all the n components failed. In our model this condition cor-responds to the time in which the execution of the *last* task end, i.e., the Fork/Join Response time. To study the system failure time we need to compute the distribu-tion function $F_{X_{max}}$ of the *maximum* X_{max} of the n variables X_i that are exponentially distributed with the same mean. Since it is $X_{max} \le t$ if and only if it is $X_i \le t$ for all $i = 1, ..., n$, it will be

$$prob[X_{max} \le t] = prob[X_1, X_2, ..., X_n \le t]$$

Thus, the distribution F_{max} of the *maximum* of n *independent* random variables X_i is given by (see, e.g., [36, 37]):

$$F_{max}(t) = prob[X_1, X_2, ..., X_n \le t] = F_{X_1}(t) \; F_{X_2}(t) \; ... \; F_{X_n}(t) \qquad (A.6)$$

Since all the variables X_i have the following identical exponential distribution with the same mean R

$$F_{X_i}(t) = prob[X_i \le t] = 1 - e^{-t/R}$$

from Eq. A.6 we have

$$F_{X_{max}}(t) = (1 - e^{-t/R})^n \qquad (A.7)$$

The distribution of Eq. A.7 is *not* exponential. To compute its mean value we may apply recursively Eq. 5.5 to a number of components decreasing from n to 1. Thus, the mean time to the first failure of n independent identically distributed compo-nents is $MTTF/n$, the following failure when the components are $n - 1$ happens after $MTTF/(n - 1)$ and so on until only one component is working that fails after $MTTF$.

System with n-Components in Series

Consider the reliability model of a system consisting of n components connected in series. In such a system we are interested in the mean time to the *first* failure, that in our model corresponds to the end of the fastest task that reach the `Join`. This equivalent problem can be stated as the computation of the distribution function $F_{X_{min}}(t)$ of the *minimum* X_{min} of n *independent* random variables X_i exponentially distributed with the same mean R. The condition $X_{min} > t$ is satisfied if and only if it is $X_i > t$ for *all* $i = 1, ..., n$. Thus, it will be (see, e.g., [36, 37]):

$$F_{X_{min}}(t) = 1 - prob\,[X_1, \; X_2, \; ..., \; X_n > t\,] = 1 - \prod_{i=1}^{n}(1 - F_{X_i}(t)) \qquad (A.8)$$

where each $F_{X_i}(t)$ is the distribution function of the exponentially distributed random variable X_i with mean R:

$$F_{X_i}(t) = 1 - e^{-t/R} \qquad (A.9)$$

Substituting the F_{X_i} in Eq. A.8 by their expressions of Eq. A.9 we obtain

$$F_{X_{min}}(t) = 1 - e^{-n\,(t/R)} \qquad (A.10)$$

As can be seen from Eq. A.10, the distribution of the *minimum* of n independent exponential distributions is itself an exponential distribution with mean $1/n$-th of the mean R of any of the individual distributions (it is much lower)!

Let us remark, that in the preceding reliability model all the n components are assumed independent and identical, with the same mean and exponential distribution. Non-repairable components are considered and no interference among events is possible (queues will never take place).

A.4   Models Described in the Book

See Table A.2.

Table A.2 Models described in the book

Subjects	Sections	Model	Class	Stations	Distributions	Tool
Web server performance	2.1	Open	Single	Source/Queue/Sink	Exp	JMVA
Capacity planning	2.2	Closed	Single	Delay/Queue	Exp	JSIMg
Use of service demands	2.3	Closed	Single	Delay/Queue	Exp	JSIMg
Optimal load of a server	2.4	Open	Single	Source/Queue/Sink	Exp	JSIMg
Multiclass workload	3.2	Closed	multi	Delay/Queue	Exp	JMVA
Optimization of data center	3.3	Closed	multi	Queue	Exp	JMVA
Variability interarr.times	4.2	Open	Single	Source/Queue/Sink	Exp/Hypo/Hyper	JSIMg
Variability service times	4.3	Open	Single	Source/Queue/Sink	Exp/Hypo/Hyper	JSIMg
Tasks synchronization	5.1	Open	Single	Source/Fork/Join/Queue/Sink	Exp	JSIMgQN
Sync. *versus* variance	5.2	Open	Single	Source/Fork/Join/Queue/Sink	Exp/Hyperexp	JSIMgQN
Sync. on fastest task	5.3	Open	Single	Source/Fork/Join/Queue/Sink	Exp/Hyperexp	JSIMgQN
Surveillance system	6.1	Open	multi	Source/ClassSwitch/Queue/Sink	Exp	JSIMg
Autoscaling fluctuations	6.2	Open	multi	Source/Queue/Place/Transition/Sink	Hyperexp/FiringImmediate	JSIMg
Workflow simulation	6.3	Open	multi	Source/Queue/ClassSwitch/Sink	Exp	JSIMg
Crowd computing	6.4	mixed	multi	Source/Queue/FCR/Sink	Exp/Hyperexp	JSIMg
Routing prob. *versus* Visits	A.1	Open	Single	Source/Queue/Sink	Exp	QN

References

1. AWS Auto Scaling (2020), https://docs.aws.amazon.com/autoscaling/index.html
2. G. Balbo, G. Serazzi, Asymptotic analysis of multiclass closed queueing networks: common bottleneck. Perform. Eval. **26**, 51–72 (1996)
3. G. Balbo, G. Serazzi, Asymptotic analysis of multiclass closed queueing networks: multiple bottlenecks. Perform. Eval. **30**(3), 115–152 (1997)
4. S. Balsamo, A. Marin, Separable solutions for Markov processes in random environments. Eur. J. Oper. Res. **229**(2), 391–403 (2013)
5. E. Barbierato, M. Gribaudo, G. Serazzi, Multiformalism models for performance engineering. Future Internet **12**(3), 50 (2020). https://doi.org/10.3390/fi12030050
6. F. Baskett, K.M. Chandy, R.R. Muntz, F.G. Palacios, Open, closed, and mixed networks of queues with different classes of customers. J. ACM **22**(2), 248–260 (1975)
7. P. Bellasi, A. Faisal, W. Fornaciari, G. Serazzi, Queueing network models for performance evaluation of ZigBee based wireless sensor networks, in *EPEW, 7th European Performance Engineering Workshop*. LNCS, vol. 6342 (Springer, 2010), pp. 147–159. ISBN 9783642157837
8. M. Bertoli, G. Casale, G. Serazzi, JMT: performance engineering tools for system modeling. ACM SIGMETRICS Perform. Eval. Rev. **36**(4), 10–15 (2009). ISSN:0163-5999. https://doi.org/10.1145/1530873.1530877
9. G. Bolch, S. Greiner, H. de Meer, K.S. Trivedi, *Queueing Networks and Markov Chains* (Wiley, 2006)
10. J.P. Buzen, Fundamental operational laws of computer system performance. Acta Inf. **7**(2), 167–182 (1976)
11. M. Calzarossa, G. Serazzi, Workload characterization: a survey. Proc. IEEE **81**(8), 1136–1150 (1993)
12. G. Casale, M. Cazzoli, S. Jiang, V.S. Lopes, G. Serazzi, and L. Zhu, Generalized synchronizations and capacity constraints for Java modelling Tools, ICPE 2017, in *Proceedings of 2017 ACM/SPEC International Conference on Perference Engineering* (2017), pp. 169–170
13. G. Casale, N. Mi, L. Cherkasova, E. Smirni, How to parameterize models with bursty workloads. ACM Perf. Evaluation Rev. **36**(2), 38–44 (2008)
14. G. Casale, N. Mi, E. Smirni, Bound analysis of closed queueing networks with workload burstiness. ACM SIGMETRICS Perf. Evaluat. Rev. 13–24 (2008)
15. G. Casale, G. Serazzi, Bottlenecks identification in multiclass queueing networks using convex polytopes. In *Proceedings of IEEE MASCOTS Symposium* (IEEE Press, 2004), pp. 223–230
16. P.J. Denning, J.P. Buzen, The operational analysis of queueing network models. ACM Comput. Surv. **10**(3), 225–261 (1978)

© The Editor(s) (if applicable) and The Author(s) 2024
G. Serazzi, *Performance Engineering*,
https://doi.org/10.1007/978-3-031-36763-2

17. L. Flatto, Two parallel queues created by arrivals with two demands II. SIAM J. Appl. Math. **45**(5), 861–878 (1985)

18. A.Ul Gias, G. Casale, M. Woodside, ATOM: model-driven autoscaling for microservices, in *Proceedings of IEEE ICDCS* (2019), pp. 1994-2004

19. A. Giessler, J. Hanle, A. Konig, E. Pade, Free buffer allocation - an investigation by simulation. Comput. Netw. **3**(1), 191–204 (1978)

20. M. Gribaudo, M. Iacono, Theory and application of multi-formalism modeling. (2013). https://doi.org/10.4018/978-1-4666-4659-9

21. M. Harchol-Balter, *Performance Modeling and Design of Computer Systems: Queueing Theory in Action* (Cambridge University Press, 2013)

22. E. Incerto, M. Tribastone, C. Trubiani, Combined vertical and horizontal autoscaling through model predictive control, in *Proceedings of European conference on parallel processing* (2018), pp. 147–159

23. L. Kleinrock, On flow control in computer networks, in *Proceedings of the IEEE International Conference on Communications - ICC*, vol. 2 (1978), pp. 27.2.1–27.2.5

24. L. Kleinrock, Power and Deterministic rules of thumb for probabilistic problems in computer communications. in *Proceedings of the IEEE International Conference on Communications - ICC* (1979), pp. 43.1.1–43.1.10

25. E.D. Lazowska, J. Zahorjan, G.S. Graham, K.C. Sevcik, *Quantitative System Performance* (Prentice-Hall, 1984)

26. J.D.C. Little, A proof of the queueing formula $L = \lambda W$. Oper. Res. **9**, 383–387 (1961)

27. Microsoft Azure autoscale (2018), https://docs.microsoft.com/en-us/azure/azure-monitor/platform/autoscale-overview

28. R. Nelson, A.N. Tantawi, Approximate analysis of fork/join synchronization in parallel queues. IEEE Trans. Comput. **37**(6), 739–743 (1988)

29. R. Pinciroli, S. Distefano, Characterization and evaluation of mobile crowdsensing performance and energy indicators. SIGMETRICS perform. Evaluat. Rev. **44**(4), 80–90 (2017)

30. C. Qu, R.N. Calheiros, R. Buyya, Auto-scaling web applications in clouds: a taxonomy and survey. *ACM Comput. Surv.* **51**(4), 73:1–73:33 (2018)

31. M. Reiser, S.S. Lavenberg, Mean-value analysis of closed multichain queueing networks. J. ACM **27**(2), 312–322 (1980)

32. A. Riska, E. Smirni, M/G/1-type Markov processes: a tutorial, in *Performance 2002*, pp. 36–63

33. F. Rossi, V. Cardellini, F. Lo Presti, Hierarchical scaling of microservices in kubernetes, in *Proceedings of the First International Conference on Autonomic Computing and Self-Organizing Systems, ACSOS 2020* (IEEE, 2020)

34. F. Rossi, V. Cardellini, F. Lo Presti, Self-adaptive threshold-based policy for microservices elasticity, in *Proceedings of Symposium on Modelling, Analysis, and Simulation of Computer and Telecommunication Systems, MASCOTS 2020* (IEEE, 2020)

35. E. Rosti, F. Schiavoni, G. Serazzi, Queueing network models with two classes of customers, in *Proceedings of the Fifth International Symposium on Modeling, Analysis, and Simulation of Computer and Telecommunication Systems, 1997. MASCOTS'97* (IEEE, 1997), pp. 229–234

36. W.J. Stewart, *Probability, Markov Chains, Queues, and Simulation* (Princeton University Press, , 2009)

37. K.S. Trivedi, *Probability and Statistics with Reliability, Queuing and Computer Science Applications* (Wiley, 2016)

38. R. Wang, G. Casale, A. Filieri, Estimating multiclass service demand distributions using Markovian arrival processes. ACM Trans. Model. Comput. Simul. **33**(1–2), 1–26 (2023)